INDUSTRIAL ENGINEERING
FOUNDATIONS

INDUSTRIAL ENGINEERING FOUNDATIONS

FOUNDATIONS

Bridging the Gap between Engineering and Management

FARROKH SASSANI, Ph.D.

The University of British Columbia

MERCURY LEARNING AND INFORMATION

Dulles, Virginia
Boston, Massachusetts
New Delhi

Publisher: David Pallai
MERCURY LEARNING AND INFORMATION
22841 Quicksilver Drive
Dulles, VA 20166
info@merclearning.com
www.merclearning.com
1-800-232-0223

F. Sassani. *Industrial Engineering Foundations.*
ISBN: 978-1-942270-86-7

Library of Congress Control Number: 2016953727

161718321   Printed in the USA on acid-free paper.

To
the loving memory of my father,
the resilience of my mother,
and
my family, from whom I borrowed the time.

CONTENTS

PREFACE

As an applied science, industrial engineering embraces concepts and theories in such fields as mathematics, statistics, social sciences, psychology, economics, management and information technology. It develops tools and techniques to design, plan, operate and control service, manufacturing, and a host of engineering and non-engineering systems. Industrial engineering is concerned with the integrated systems of people, materials, machinery, and logistics, and ensures that these systems operate optimally and efficiently, saving time, energy, and capital. An industrial engineer, or an engineer equipped with the tools and techniques of industrial engineering is the professional who is responsible for all these tasks.

This book aims to expose the reader at an introductory level to the basic concepts of a range of topics in industrial engineering and to demonstrate how and why the application of such concepts are effective.

The target audience for this abridged volume encompasses all engineers. In other words, the book is written for motivated individuals from a broad range of engineering disciplines who aim for personal and professional development. They would benefit from having a foundational book on important principles and tools of industrial engineering. They would be able to apply these principles not only to initiate improvements in their place of work but also to open up a career path to management and positions with a higher level of responsibility and decision-making.

The level of coverage and the topics included in this book have been distilled from over thirty years of teaching a technical elective course in industrial engineering to non-industrial engineering students in their final year of studies. From direct and indirect feedback from the students on the usefulness of various industrial engineering methods in their work, the course content has evolved to what is now presented as a book.

In writing this book, I am most grateful to Dr. Mehrzad Tabatabian, from the British Columbia Institute of Technology, for his encouragement, sharing his book-writing experiences, and putting me in contact with a wonderful publisher, Mr. David Pallai, who supported and provided guidance throughout the effort. I am also indebted to my colleagues in the Mechanical Engineering Department at the University of British Columbia: Professor Peter Ostafichuk and Mr. Markus Fengler for giving me invaluable advice about writing a book, and I deeply acknowledge a special debt of gratitude to Professor Tatiana Teslenko for graciously editing and proofreading the manuscript.

This is an opportune time to thank Ms. Aleteia Greenwood, Head of UBC Woodward Library and Hospital Branch Libraries, for her advice and expert assistance in the literature search and library matters for many years. I also would like to extend my appreciation to my graduate students Morteza Taiebat, Abbas Hosseini, and Masoud Hejazi for their tolerance. Last, but not the least, I express my heartfelt gratitude to my family for their patience, encouragement, and support.

F. Sassani
November 2016

Chapter 1 | INTRODUCTION

1.1 INDUSTRIAL ENGINEERING

A complete study of industrial engineering would typically require four years of university education. Industrial engineering sits at the intersection of many disciplines, such as engineering science and technology, social sciences, economics, mathematics, management, statistics, and physical sciences. An industrial engineer is akin to a manager on the site of operations. Industrial engineering is an ideal bridge between all hard-core engineering fields such as mechanical, chemical, electrical, and manufacturing engineering, and the field of management. This book is an introduction to the basic principles and techniques used by industrial engineers to control and operate production plants and service systems. The reader will become familiar with the most important industrial engineering concepts and will gain adequate knowledge of where and what to look for if he or she needs to know more.

Some engineers, in their own words, are unenthusiastic and unexcited about their long-term careers in engineering because they envision no possibility for enhancing their effectiveness or progress through the ranks. Many would like to be involved in management, process improvement, and decision making and in this way pave a path to production control, engineering and its management, and the operational aspects of manufacturing and the service industries in which they work. It is not always practical to earn another relevant degree. Without knowledge and understanding of how an enterprise operates, and how to improve the efficiency of tasks and processes, it will be difficult to become involved, or be called on to participate in the management of engineering operations and decision making. We witness how various disciplines, such as mechanical engineering and electrical engineering, have joined forces to train "Mechatronics Engineers" or biomedical science and engineering have pooled together their expertise to train "Biomedical Engineers." Industrial engineering is a discipline that, when integrated with any other engineering field, can educate competent engineers who not only design and develop equipment, processes, and systems but also know how efficiently utilize and manage them.

An industrial engineer attempts to make an organization more efficient, cost-effective, and lean, and he/she may deal with topics like "productivity" and "profitability." For the sake of an example, let's briefly discuss the two and see what we learn.

$$Productivity = \frac{Output}{Input} \; such \; as \; \left[\frac{\$ \; profit}{Number \; of \; staff} \right] or \left[\frac{Units \; produced}{Time \; period} \right]$$

$$Profitability = Revenue \; from \; Sales - Cost \; of \; input \; [\$]$$

It is expected that a productive firm will also be a profitable one. However, because profits are strictly sales oriented, there must be sufficient sales to be profitable. A manufacturing enterprise may be very efficient in the use of its resources, but reviews for inferior quality, missed delivery promises, and poor customer service, for example, may hinder sales to secure a net profit. Therefore, firms have to not only deal with many internal factors, but they also have to consider the external influences ranging from raw material availability and cost of purchase to clientele opinion, union matters, and governmental regulations.

1.2 DUTIES OF AN INDUSTRIAL ENGINEER

Industrial engineers are hands-on managers who determine, eliminate, enhance, or implement, as appropriate, to deal with the following:

- Waste and deficiencies
- Better means of control
- Cost-effective analysis
- Improved technology
- More efficient operations
- Greater use of plant facilities

- Safety and human aspects
- Equipment layout
- Delivery and quality issues
- Personnel assignment
- Other related functions and tasks depending on the case

There are many techniques and principles available to industrial engineers to effectively tackle and manage any of the above scenarios.

1.3 SUBJECT COVERAGE

The topics covered in this book comprise some of the important concepts and tools of industrial engineering and are certainly not exhaustive by any means. Some subjects are still of great importance in the field and in practice, such as inventory control, but more and more inventory controls are implemented through commercial software that is readily available and widely used. For such reasons and the fact that this book is intended to cover the foundations of industrial engineering, essentially, for nonindustrial engineers, the subject coverage converged to what is presented. Figure 1.1 shows the topics and what is expected to be achieved if they are applied to any enterprise of interest.

1.4 SUGGESTIONS FOR READING THE BOOK

The goal of this book is to provide an overview of some important concepts in industrial engineering. It is also intended to be light and portable and, thus, readily at hand for reading practically at any time. Every reader has his or her own style of reading that suits his or her available time, learning habits, and specific subjects of interest. If the intention is to become familiar with and understand the foundations of industrial engineering, we suggest reading the book in its entirety while also going through the examples. In the first attempt, it is not necessary to engage too much with the derivations and the exercises. Once the overall ideas have been grasped, then subjects of greater relevance to one's work can be studied in detail.

1.5 HOW TO BE A MANAGER

One purpose of this book is to assist engineers to take on managing responsibilities. Whereas engineers are known to be logical and systematic, and they must be precise in

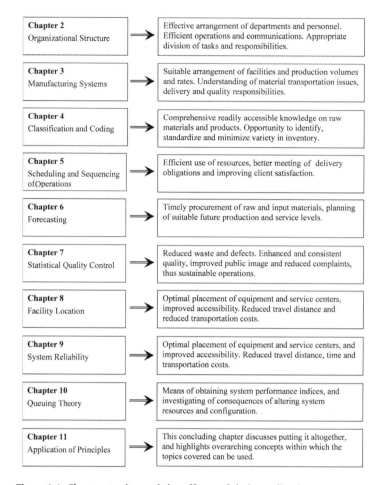

Chapter 2 Organizational Structure	Effective arrangement of departments and personnel. Efficient operations and communications. Appropriate division of tasks and responsibilities.
Chapter 3 Manufacturing Systems	Suitable arrangement of facilities and production volumes and rates. Understanding of material transportation issues, delivery and quality responsibilities.
Chapter 4 Classification and Coding	Comprehensive readily accessible knowledge on raw materials and products. Opportunity to identify, standardize and minimize variety in inventory.
Chapter 5 Scheduling and Sequencing of Operations	Efficient use of resources, better meeting of delivery obligations and improving client satisfaction.
Chapter 6 Forecasting	Timely procurement of raw and input materials, planning of suitable future production and service levels.
Chapter 7 Statistical Quality Control	Reduced waste and defects. Enhanced and consistent quality, improved public image and reduced complaints, thus sustainable operations.
Chapter 8 Facility Location	Optimal placement of equipment and service centers, improved accessibility. Reduced travel distance and reduced transportation costs.
Chapter 9 System Reliability	Optimal placement of equipment and service centers, and improved accessibility. Reduced travel distance, time and transportation costs.
Chapter 10 Queuing Theory	Means of obtaining system performance indices, and investigating of consequences of altering system resources and configuration.
Chapter 11 Application of Principles	This concluding chapter discusses putting it altogether, and highlights overarching concepts within which the topics covered can be used.

Figure 1.1. Chapter topics and the effects of their application to systems.

what they do, they generally are not considered as good managers. This may be true for those engineers who have moved up through the ranks in engineering design. Unless engaged with a group on a large-scale project, design is a solitary profession where individuals work on their own, solely responsible for the design of a component or a system. A designer likes to concentrate without being interrupted.

The majority prefer it this way, quietly accepting the challenges of engineering design as individuals who have a high regard for their own design skills—more power to them, because the society and the engineering community need dedicated design engineers.

Most of these engineers choose the design of products in distinct preference to the uncertainty of working with people. One reason might be that their academic backgrounds generally did not include training in directing and motivating other people; they concentrated on becoming technically competent. Thus, they usually find it difficult to make the transition from dealing with the technical discipline to dealing with people. Industrial engineers, however, must work with many individuals, and, as mentioned earlier, they are managers of systems and people on the site. They must arrange one-to-one and group meetings; seek input, feedback, and suggestions for improvement; and act on them by using industrial engineering methods and tools.

1.5.1 Guidelines to Follow

There are a number of guidelines to effective management that should be followed when an engineer transfers from his or her independent design role to the additional job of manager. These guidelines can help one organize and encourage employees to greater productivity.

The first guideline is to establish personal relationships. Such relationships should be comfortable and open. Engineers have justifiable pride in their professional skills and want to be treated by their managers with respect and as equals. They also want to know what is going on and feel they have the right to ask questions in order to get such information.

While managers clearly cannot be fully open to their employees on confidential matters, they should make every effort to keep areas of confidentiality to a minimum. Those instances where they cannot be fully open to their employees should be the exceptions to their practice.

A manager should also be systematic and organized. Managers who are organized will help their employees see their own tasks more clearly and thus be organized and systematic in their work. Much of this organization is observed in the information a manager conveys to the employees about their jobs and the way a manager uses his or her own time.

1.5.2 Working with People

Employees need to know the relationship of the parts to the whole. They may have many questions about the roles of other people in the organization, and how their jobs relate to others. They need to be clear on lines of authority and know the extent of their own authority. This subject matter is addressed in Section 2.2.

It is important to assess the work of employees with an emphasis on learning. Managers should not comment only if something goes wrong. When this occurs, employees hear the comments as a criticism of them personally. If comments and observations are both positive and negative, it is much more likely that employees will interpret them as constructive and learn from the process. For example,

Chapters 5 and 7 cover topics that require feedback and data from the personnel, as well as instructions to be given to them. Their cooperation is certain when the changes do not imply criticism of their work.

Managers ought to be on the lookout and have an attitude of readiness to change. As new ideas emerge and technology advances, earlier concepts become obsolete. Therefore, it is important for managers to have an open mind and be willing to listen to employees' ideas and suggestions. There is a good chance that these ideas will lead to a better way of doing things.

Employees can sense a manager's willingness to listen to their ideas. By creating a receptive and dynamic climate, employees will be less hesitant to express their creative and innovative ideas to perform the tasks better and to become more productive.

An effective manager will delegate as much as possible, but no more! There are two major reasons why managers do not delegate well. First, they are skilled in the technology or the tasks themselves and frequently think others cannot perform the work as well as they do. Second, they do not know how to teach others to be as good as they are. They often perform their work intuitively and are not able to explain how it is done. To be a good manager, one will need to overcome these two barriers to delegation if productivity or quality of service is to improve.

1.5.3 Decision Making

A manager must also know how frequently to give individual employees information on how they are doing. Some people need feedback more frequently than others. When managers make a decision, it usually affects all the personnel working for them. Thus, they will want to feel they had a say in the decisions that affected them.

Consultative decision making has three positive results. First, it teaches employees an analytical approach to the process. Second, it helps employees feel more involved in the important issues affecting them. Third, it encourages employees to be committed to carrying out the decisions reached. As a manager, in every chapter of this book, you will find an opportunity to seek feedback. As an engineer, knowing the subject matters in this book, you will have the courage to share your ideas, express your opinion, and be in a position to participate in decision making and management functions.

Chapter 2 ORGANIZATIONAL STRUCTURE

2.1 INTRODUCTION

An organization is an association of individuals supported by resources and capital working collectively to achieve a set of goals and objectives. There are many factors in effectively managing an organization, and one of the most essential is the organizational structure. Without a structure that defines the flow of information and physical entities, control of operations, and the tasks and responsibilities, it is doubtful that anything will function efficiently and as desired. It is then clear that the goals of any nature will be difficult to achieve.* Considering a manufacturing company as an example, Figure 2.1 shows the diversity of the functions and entities that could be involved.

Referring to Figure 2.1, it is hard to imagine how an enterprise can operate without thoughtfully bridging so many islands of activities and prescribing appropriate rules of conduct. Organizational structure and management are so important that it is essential that when even two individuals cooperate in a work effort, the tasks are divided, but only one of them directs the activities of the team. Otherwise, conflicting situations will arise. Therefore, every enterprise must have a well-integrated

---

* We frequently refer to production and manufacturing environments to exemplify an organization that reflects the engineering scope of this book and to allow us to highlight the key concepts better. In general, these concepts can be equally applied to many organizations that have identifiable entities and functions, and provide products or services.

organizational structure for its personnel and departments, the form of which depends on the goals of the enterprise. A consulting firm, a hospital, or a chemical plant, for example, all have different organizational needs and structures. In a manufacturing or production company, the term *production* implies transformational processes in which raw materials are converted into consumable goods and products. In doing so, the combination of manpower, machinery, tools, energy, and capital is required. This is a simplified definition. As depicted in Figure 2.1, there are many tasks, such as financial planning, demand forecasting, operations scheduling, quality control, and distribution, that are associated with manufacturing any product. This association is such that, if satisfactory performance is required, which is undoubtedly always the case, then there is a need for well-designed structure to control and coordinate all these highly interrelated activities. In fact, no human undertaking is ever successful unless it is organized. Therefore, to achieve the goals of an organization, it is necessary to develop teams or units that function as a single efficient and effective entity.

Figure 2.1. Entities and functions in a manufacturing company.

2.2 BASIC PRINCIPLES OF ORGANIZATION

We discuss various forms of organizational structure later; however, regardless of these forms, there are basic operational principles that are common to all. At the personnel level, all members of an organization perform better in their roles when they know the following clearly:

- What their job is
- Who their direct supervisor is
- Where their position in the organizational structure is
- What their authority is
- To whom they must report relevant information

These factors are essential to a good organization and apply to everyone in the organization. It is fundamental for every position that there is a clear job description, and the members pay attention to and understand their roles and responsibilities, both individually and collectively. A key rule is that every member should report to only one supervisor. Knowing their position within the organizational structure enables the personnel to follow the proper channels of interaction, patterns of information and material flow, and appropriate forms of communication. On occasion, unanticipated situations, such as an emergency, can arise that may require immediate action. Despite the fact that no rules can be set out ahead of time for unforeseen events, some guidelines as to who among the personnel present must act, and on what degree of authority, will be useful for all staff.

At the organizational level, it is clear that no universal principles can be developed because different organizations require different forms of structure. However, the following general guidelines apply in many instances:

- **Clear Objectives:** The appropriate form of an organization depends directly on what is to be achieved. The organization should have clearly stated objectives, and every unit of the organization should understand and work toward achieving its purpose.
- **Logical Work Division:** An organization must be divided into appropriate units where work of a similar nature is performed.
- **Unitary Command:** Every organization should have only one person at the top who is responsible for its performance.
- **Work Delegation:** It is impractical for one person to directly supervise and coordinate the work of too many subordinates. Therefore, top persons must assign duties to capable associates and subordinates within the framework of the organization.
- **Well-Defined Responsibilities:** Subordinates must be given a job description and some autonomy to freely perform their jobs. All responsibilities assigned to all units and all members should be clear, specific, and understandable.
- **Reciprocative Communications:** Top persons should also seek feedback from subordinates in improving processes and participating in decision making.

Many organizations develop vision and mission statements to set out their objectives. A vision statement expresses an organization's ultimate goal and the reason for its existence. A mission statement provides an overview of how the organization plans to achieve its vision. Vision and mission statements are well publicized as identifying symbols of the organization.

Without underestimating the importance and necessity of developing a vision statement, a comparison of such statements reveals significant similarities. This is particularly noticeable in certain areas with closely related goals. For example, how much can a vision statement of one university differ from another? Nevertheless, universities find a particular niche on which to build. Therefore, any organization must spend some effort in developing a vision statement that succinctly reflects their philosophy for the products or services they offer. For the substantial detail on its vision and mission statements, we consider the University of British Columbia (UBC) as an illustrative organization. The following is the vision statement of UBC.

Vision:

As one of the world's leading universities, The University of British Columbia creates an exceptional learning environment that fosters global citizenship, advances a civil and sustainable society, and supports outstanding research to serve the people of British Columbia, Canada, and the world.

Mission statements are sometimes expressed in several annotations. UBC states these as values:

Values:

Academic Freedom

The University is independent and cherishes and defends free inquiry and scholarly responsibility.

Advancing and Sharing Knowledge

The University supports scholarly pursuits that contribute to knowledge and understanding within and across disciplines and seeks every opportunity to share them broadly.

Excellence

The University, through its students, faculty, staff, and alumni, strives for excellence and educates students to the highest standards.

Integrity

The University acts with integrity, fulfilling promises and ensuring open and respectful relationships.

Mutual Respect and Equity

The University values and respects all members of its communities, each of whom individually and collaboratively makes a contribution to create, strengthen, and enrich our learning environment.

Public Interest

The University embodies the highest standards of service and stewardship of resources and works within the wider community to enhance societal good.

Within a large organization, often individual units have their own vision and mission statements, but under the umbrella of the organization's core statements. This is more so for the mission statement because its purpose is to elucidate the commitments and means of accomplishing the vision. Since different units in an organization have difference ways and tools, it is appropriate to have a suitably tailored mission statement.

The following are two examples of the mission statement from two different units at UBC.

Mission:

UBC Continuing Studies is an academic unit that inspires curiosity, develops ingenuity, stimulates dialogue and facilitates change among lifelong learners locally and internationally. We anticipate and respond to emerging learner needs and broaden access to UBC by offering innovative educational programs that advance our students' careers, enrich their lives and inform their role in a civil and sustainable society.

Mission:

UBC Library advances research, learning and teaching excellence by connecting communities, within and beyond the University, to the world's knowledge.

2.3 FORMS OF ORGANIZATIONAL STRUCTURE

There are no rules or uniquely defined models for organizational structure. For a given company and known set of objectives, and depending on the philosophy taken, the specific organizational structure will emerge. The basic structure of an industrial organization depends on the size of the company, the nature of business, and the complexity of its activities. However, there are some basic forms of structuring an organization. The most common is the "line and staff" structure, but there are a number of variations of this basic form that will be reviewed in the following sections. The layouts of these forms are referred to as the organization or organizational chart.

2.3.1 Line Organization

The simplest form of organization is the straight line. The straight-line concept implies that the positions on the organizational chart follow a vertical line, as shown in Figure 2.2.

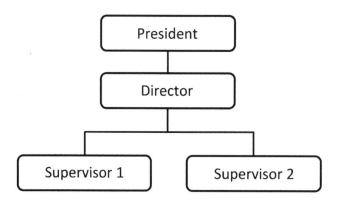

Figure 2.2. **Line Organizational Chart.**

In such an organization, the chief or president handles all issues that arise, from production to sales, finance, or personnel, because the chief or president has no specialized staff to aid them. Small and start-up companies begin with the line type of organization. The principal owner is the general manager or the president, and he or she will possibly have only a few staff members working in the company. As the business grows, he or she will be overwhelmed with diverse tasks, and will need to expand the organization by upgrading to a "line and staff" type of structure.

2.3.2 Line and Staff Organization

The line and staff organization in a simplified form is shown in Figure 2.3.

The main difference between this and the line type of organization is the addition of some specialists who are appointed where the president needs the most assistance. The chief engineer may be responsible for both design and manufacturing processes. The sales manager, in addition to sales, may also be responsible for marketing and public relations. Certain responsibilities could be shared by two specialists, but they report to the president, who retains the overall responsibility for the operation of the enterprise.

The addition of staff and specialists can be extended depending on the growth and size of the organization and the diversity of the activities involved. Figure 2.4 illustrates a typical organizational structure of a medium-size manufacturing company.

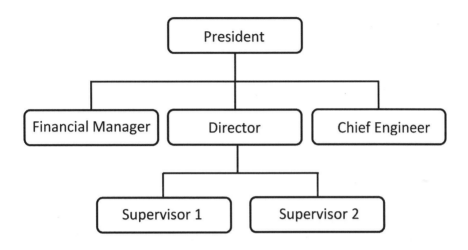

Figure 2.3. **Line and Staff Organizational Chart.**

For larger companies it will be difficult to adhere to a pure line or even line and staff type of organization, which normally expand horizontally. Owing to the amount of work and the diversity of activities, and in order to maintain efficiency and productivity, it is sometimes necessary to incorporate vertical levels of authority and work division. This, however, should be done judiciously, since a large number of organization levels can impede communications and adversely affect the efficiency of operations. In very large organizations, ten or more vertical levels of structuring may be seen, but in such instances, essential operational information does not have to pass through all the levels. In any case, despite the need, the number of vertical levels should ideally not exceed five.

2.4 ORGANIZATIONAL DESIGN STRATEGIES

The design of an organization depends on many factors, and its structure at any time will reflect those factors. Important factors in the structure of an organization are the following:

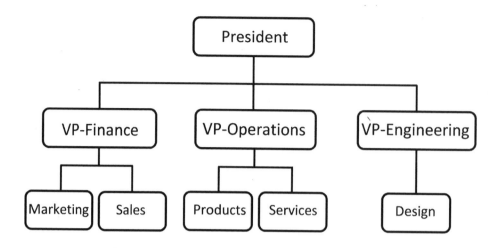

Figure 2.4. Organizational chart for a manufacturing company.

- Size of the enterprise in terms of personnel, resources, and capital
- Nature of the primary activities to be coordinated, such as those specific to manufacturing, banking, or health care
- Issues concerning human resources, such as qualification, the level of authority, and salary or commission
- Philosophy of the management, such as gaining a good public image, earning maximum profit, or providing the best service
- Communication patterns necessary to coordinate all the operations in the best possible manner, such issues as verbal and written protocols, handling of confidential data, and record-keeping matters

The structure of an organization evolves from a series of design activities known as "departmentation." This means the grouping of specialized activities and resources to accomplish a total task.

2.5 MAJOR FORMS OF DEPARTMENTATION

Theoretically, there are as many ways to departmentalize as there are organizations. Even two organizations of similar products and goals can differ in details as a result of many reasons, from geographical location to management style. However, significant generic forms of departmentation are as follows:

- Functional
- Product/divisional
- Geographic
- Clientele
- Process
- Time
- Alphanumeric
- Autonomy

A description of these concepts for departmentation and their important merits/demerits is presented next.

2.5.1 Functional Departmentation

This most common type of departmentation is ideal for small-medium companies, where the organization is divided into departments according to the tasks they must perform. There is typically only one of each specialized unit, such as finance, manufacturing, marketing, and so on, and normally in one location, but not necessarily. This formation is especially suitable for single-product companies, such as steel foundries and aluminum plants.

Advantages

- Efficient operations result from specialization in each department.
- Focusing on specific related tasks provides an opportunity for continuous improvement in personnel and organizational performance.
- Effective long-range planning and major decision making are more readily accomplished.

Disadvantages

- After an optimal size is reached, poor communication and coordination often ensue.
- Lack of motivation and a feeling of being unimportant can develop.
- Overlapping authorities and problems with decision making can appear.

2.5.2 Product/Divisional Departmentation

A second common form of departmentation is organized along the product lines. Most companies organized in this fashion were initially functionally organized. Expanding product diversity can overwhelm the organization in every aspect of its operation and can cause inefficiencies. In product departmentation, different products or family of products are managed separately. For example, there may be divisions for audiovisual equipment, appliances, and health-care products each having their own finance, production, marketing, and other relevant departments.

Advantages

- It efficiently integrates the main activities required to make a particular product.
- It works independently while relying on support from the mother organization.
- It allows development of significant expertise as divisions compete to supersede in performance.

Disadvantages

- It impedes the direct transfer of information and technical knowledge between divisions.
- It necessitates duplication of some resources.
- Recruiting and retaining capable managers requires effort and can be costly.

2.5.3 Geographic Departmentation

The purpose of this form of departmentation is to provide goods and services in the geographical locations where they are required, such as post offices, hospitals, and transit systems that serve communities at dispersed geographical locations. Large corporations can have departments in various cities or countries.

Advantages

- There is effective provision of goods and services closer to the location of demand.
- It provides the opportunity to hire local personnel.

Disadvantages

- There can be duplication of resources.
- The possibility exists for a lack of direct management and support.

2.5.4 Clientele Departmentation

Customer or clientele departmentation is used to provide specific goods and services for specific classes of clients, such as children, youth, or adults.

Advantages

- Specialization and expertise in products and services are possible.
- Better relationships with clients can be fostered.

Disadvantages

- There can be a duplication of resources.
- There may be difficulty in maintaining uniform services.

2.5.5 Process Departmentation

This form of departmentation is common in some manufacturing plants, where processes of similar nature are grouped together to enhance resource availability and accumulate a greater level of expertise.

Advantages

- It improves coordination of similar processes.
- It allows better decision making and implementation of changes.
- It reduces duplication of resources.

Disadvantages

- It lacks coordination with other processes.
- It creates conflict in authority in large departments.

2.5.6 Time Departmentation

This form of departmentation involves the breaking down of the organization's operations into time-dependent components. Often, resources must be reorganized and reassigned to accomplish the required tasks within the allotted time.

Advantage

- The organization can be flexible and dynamic in responding and adapting to change.

Disadvantage

- There is difficulty in establishing permanent and consistent procedures.

2.5.7 Alphanumeric Departmentation

Alphabetical and numerical departmentations are used to assign employees to a team or department when the level of skills required is not significant, or there is evidence that skills needed are relatively homogeneous. It can also be used to divide the work, such as postal code for distribution of mail, or direct the clientele, such as service counters in a passport office.

Advantage

- There is great ease and speed of implementation.

Disadvantage

- If the assumption of homogeneous skills is unrealistic, it can result in departments with imbalanced capabilities.

2.5.8 Autonomy Departmentation

If the activities of an organization are diverse, unrelated to one another, and do not require "direct" management, then this form of departmentation is the most appropriate.

For instance, major conglomerate corporations, whose main goal is to invest in well-established companies regardless of the products or services offered, fall in this category. They show little interest in directly managing the acquired companies and allow autonomy in their operations.

Advantage

- There is freedom from direct involvement.

Disadvantage

- There can be a lack of knowledge about developing problems.

2.6 MANAGEMENT FUNCTIONS

One of the many purposes of any organizational structure of an enterprise is to enable and guide the management in its operations. These pertain to specific functions of planning, organizing, directing, and controlling.

- *Planning* is the process of determining the future activities by developing goals and objectives, and establishing guidelines and timelines for achieving them.
- *Organizing* deals with providing and integrating the required resources, such as personnel, equipment, materials, tools, and capital to perform the planned activities.
- *Directing* assigns specific tasks and responsibilities to personnel, provides support, motivates, and coordinates personnel's effort.

- *Controlling*, which is more associated with directing, evaluates performance, rectifies issues, and ascertains that all activities are satisfactorily progressing toward the set goals and objectives.

Figure 2.5 shows the process flow within these four functions.

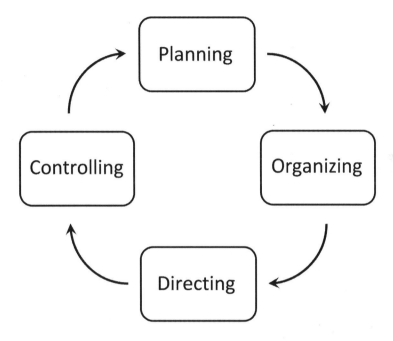

Figure 2.5. **Management functions.**

The management functions are dynamic in the sense that they apply to ongoing activities within, and they also respond to changes occurring outside the organization, such as new technology, regulations, and economic conditions. For instance, if it becomes mandatory to implement certain quality assurance plans to improve safety, the entire organization will be affected, and the functions of planning, organizing, directing, and controlling will be responsible for accomplishing the change.

2.6.1 Contemporary Operations Management

There are ever-increasing trends to provide high-quality goods and services. Whereas traditional operations management functions ascertain that ongoing activities are proceeding satisfactorily, contemporary operations management attempts to be more dynamic and improve the nature and effectiveness of the management functions. Activities of a progressive forward-looking organization include the following:

- Gathering information on customer needs, gaging their satisfaction, and using these to improve the products and services offered
- Exploiting new technology to improve quality, productivity, and efficiency
- Providing an opportunity for all members of the organization for professional and skill development
- Genuinely linking with customers and community to improve the well-being of society

2.6.2 Span of Control

Span of control refers to the number of personnel who report to one supervisor. It is one of the most widely known and used organizational principles. The ideal span of control is between four and eight. In practice, however, there is a much greater span than this, perhaps twelve or more—but this is undesirable. Higher levels in organizations have a smaller span of control. A typical company or university will have one president and possibly five or so vice presidents for various functions, such as finance, public relations, research and development, and operations.

2.6.3 Complexities of Span of Control

With the increase in the span of control, the different combinations of the relationship with subgroups increase exponentially. There are mathematical equations to determine the total number of such relationships, but it will be more tangible if we show this by means of an example. Suppose there is one supervisor S, and there are five subordinates A, B, C, D, and E.

Excluding a large number of possible relationships among the subordinates, the number of relationships between the *supervisor* and the *subordinates* is still high, as shown in Table 2.1.

Clearly, this is an extreme representation, but it is easy to see how the combinations can rise dramatically. With five subordinates, if we assume many relationships unnecessary in practice, even a dozen different relationships are difficult to maintain. It is also very challenging to arrange a common time for meetings when every subordinate is available.

2.6.4 The Informal Organization

The discussion in the preceding sections concerned what might be termed the *formal organization*. We showed precise relationships among the various positions

Table 2.1 Relationships between one supervisor and five subordinates

Pattern of Relationships Involving Supervisor	Number of Relationships
With listed individual when the other four are not present: S → A, B, C, D, E	5
With listed two individuals when the other three are not present: S → AB, AC, AD, AE, BC, BD, BE, CD, CE, DE	10
With listed three individuals when the other two are not present: S → ABC, ABD, ABE, ACD, ACE, ADE, BCD, BCE, BDE, CDE	10
With listed four, when the remaining individual is not present: S → ABCD, ABCE, ABDE, ACDE, BCDE	5
When all individuals are present: S → ABCDE	1
Total relationships	31

in the enterprise in terms of a formal organizational chart, and the need for such an organization is emphasized. We must also consider the facts of life within an organization, however, and recognize that these positions are held by people with social interests. As a result, some informal relationships will always exist. Superimposed on the formal organizational chart in Figure 2.6 is an informal organization that might exist. The informal organization is shown by the dashed lines.

Executive 2b, for instance, has some relationship with employee h4, with whom he or she has no formal organizational connection. Such informal relationships exist for a variety of reasons. Some people are natural leaders and problem solvers wherever they are placed in an organization, and other people will go to them for advice regardless of the formal organizational structure. Friendship within and outside the organization is another possibility that can manifest itself as an inevitable informal organization. Informal relationships could mean disruption of formal channels, but they could also have positive effects depending on the nature of the relationships and the conduct of activities. Horizontal communications are enhanced by informal organization, thereby improving cooperation and coordination between departments. If all communications had to be vertical and conform strictly to an organizational chart, efficiency would soon diminish.

With the advent of social media, particularly the ease with which communication occurs via texting, there will be a greater degree of networking, relationships and, thus, an informal organization with a much higher level of personal requests, favors or urgent action. Whereas these activities can lubricate the functional machinery of the organization, they can also severely violate privacy and personal rights and eventually endanger the organization's well-being. The informal organization can neither be encouraged nor completely prohibited. A healthy level of informal organization will actually be beneficial. Organizations, however, must be aware of this; set reasonable, well-intended, and practical

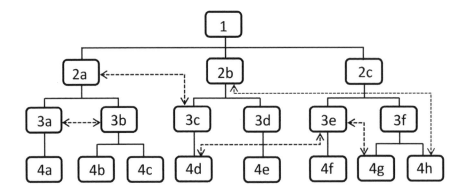

Figure 2.6. Informal relationships in organization.

guidelines; and inform all personnel of the consequences of an out-of-control interpersonal network.

2.7 ORGANIZATIONAL ACTIVITIES OF A PRODUCTION CONTROL SYSTEM

In the previous sections, the overall organizational structure for production was reviewed. At the operational and control level, a different form of structure showing the relationships between the various functional elements can be constructed. Figure 2.7 illustrates the management functions and decisions necessary to plan and control production and classifies them by the length of the planning horizon needed to consider adequate factors that are relevant to the decision.

Long-range decisions involve the definition of priorities, product lines, the establishment of customer service policies, selection of distribution channels, further investment, determination of warehouse capacity, and perhaps allocation of capacity to different product lines. The decisions are made and reviewed quarterly or annually and require planning horizons of one to five years. Market research, long-range forecasting, and resource planning are necessary activities.

Intermediate-range planning is done within the framework of policy and resource constraints resulting from the long-range planning process. Planning horizons of three months to a year are often sufficient. Necessary management functions include forecasting, workforce planning, and production planning. Typically these activities are carried out on a monthly cycle, although it is not unusual to see a shorter review period used.

Short-range activities involve scheduling and control of production. Decisions involve adjustment of production rates to adapt to forecast errors, material shortages, machine breakdown, and other uncertainties; the assignment of workers to work activities; the determination of priorities and the sequence of operations; the assignment of work to workstations; the use of overtime; and the adjustment of

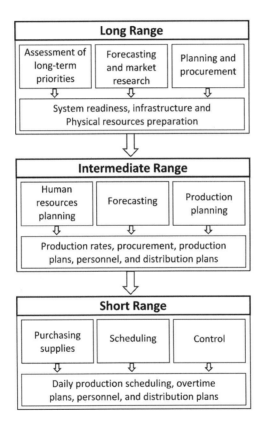

Figure 2.7. Planning horizons for production control.

in-process inventory levels. Formally these activities may be carried out weekly or daily, but they often are a continuous function of the production control department and line supervision. Decisions can be made at any time. The planning horizon is short, usually one or two weeks, but could be longer if the manufacturing cycle time is long.

2.8 TYPICAL PRODUCTION AND SERVICE SYSTEMS

Figures 2.8 and 2.9 represent arrangements for a production system and a service system. Such arrangements as shown in Figures 2.8 and 2.9 are referred to as the "manufacturing cycle," or the "operational cycle," since the information flow is more or less cyclic through the system. The arrangements shown in these figures are not unique, and variations are possible depending on the nature of the products or services, the size of the enterprise, specific objectives, and the management style. In both figures, the department designated as "finance" has been distinctly identified to show that every aspect of the organization's activities has financial implications.

Despite different appearances of the structure of production and service system as depicted in Figures 2.8 and 2.9, both share similar functional activities, the pattern of information flow, and application of industrial engineering techniques. Basically, they bridge together many of the islands of activities shown in Figure 2.1.

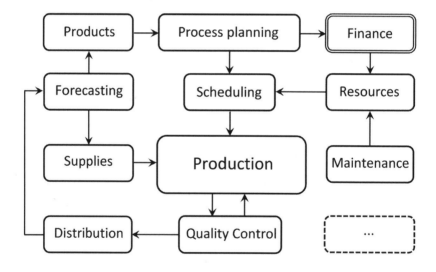

Figure 2.8. **A production planning and operation system.**

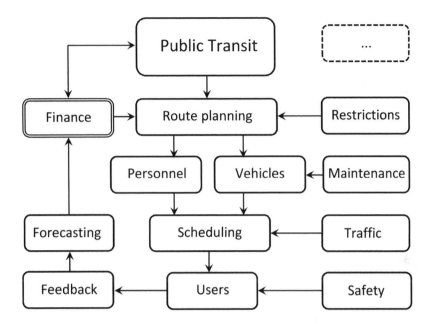

Figure 2.9. **A service planning and operation system.**

EXERCISES

2-1   Referring to Figure 2.1:

  a.   Identify the functions and entities that are relevant to your current place of occupation.

  b.   List functions and activities that your present work deals with, but that do not appear in Figure 2.1.

  c.   Make a list of all the functions relevant to your present work. Have items of the list in mind while reading the book, and note where each topic may be applicable.

2-2   Draw an organizational chart for an enterprise that you have worked for and do the following:

  a.   Highlight your position on the chart.

  b.   Examine the presence of what we called the "informal organization" and superimpose it on the chart.

    c. If you sensed that some sort of span of control was in place, what were the typical numbers?

    d. Can you say anything especially positive or negative about your organization's structure?

2-3 Using the Internet, search and obtain an organizational chart for a range of companies, industries, or governmental agencies to see how they organize themselves to offer better services and achieve their objectives.

2-4 Think of an enterprise that may be best organized by geographic departmentation, draw a representative organizational chart, and write a vision statement for it.

2-5 Find a company that may be best organized by clientele departmentation, draw a representative organizational chart, and write a mission statement for it.

2-6 In your place of work, can you identify individuals, groups, or departments who deal with planning, organizing, directing, and/or controlling?

2-7 In your place of work, identify the activities that fit long-range, intermediate-range, and short-range planning.

Chapter 3 MANUFACTURING SYSTEMS

3.1 INTRODUCTION

The operation and performance of a manufacturing system depend heavily on its physical layout, and this physical layout varies significantly from one firm to the next. The ability to recognize the dominant form of a manufacturing system of an enterprise and its layout is an important first step for analyzing, assessing, and applying appropriate controls and improvement plans.

The nature of a company's manufacturing system depends on many factors. The size of the firm, the type of products, production volume, and the diversity of the products influence the adoption of a particular organization for its manufacturing resources.

Over the last several decades, owing to significant advances in automation, robotics, sensing technology, and computer science, manufacturing systems have been revolutionized and fully automated, flexible, and many forms of computer-integrated manufacturing plants have emerged. Nevertheless, a significant proportion of products is still manufactured in the conventional form, although more and more older machinery is being replaced with advanced precision equipment. Being familiar with both traditional and modern manufacturing systems is essential for industrial engineers.

3.2 CONVENTIONAL MANUFACTURING SYSTEMS

In general, there are three types of conventional manufacturing systems:

1. Job shop production
2. Batch production
3. Mass production

Job shop–type environments will probably always exist in the classical form, as will be discussed in Section 3.2.1. Batch manufacturing systems encompass well over 50 percent of the world production of goods, and, along with mass production systems, they can be highly automated and "modern" in equipment. Next, we review these manufacturing systems. Modern concepts of manufacturing systems are discussed in Section 3.4.

3.2.1 Job Shop Production

This most general form of manufacturing, also called *one-off*, is characterized by production volumes of small quantity that are often nonrepeating. Therefore, specialized and automated types of machinery and equipment are usually not required. Job shop production is used to handle a large variation in the type of products and work that must be performed. This is the reason that production equipment must be general-purpose to allow for the variety of work, but the skill level of the personnel must be relatively high so that they can perform a range of nonroutine work. Successful operation in a job shop environment relies heavily on the skills of its workforce and the flexibility in adapting to the changing conditions. Products typically manufactured in a job shop are aircraft, space vehicles, prototypes of new products, and research and experimental equipment.

Shipbuilding and fabrication of major outdoor structures are not normally categorized as job shop production, although the quantities produced are typical of a job shop environment.

Often with one-of-a-kind production and no previous experience the production rate of a job shop will be low compared with other manufacturing systems. This does not necessarily mean that the job shop environment is inefficient or unproductive. In fact, many products that are produced in higher quantities and rates in dedicated systems were initially manufactured in a job shop environment as prototypes. Production in a job shop environment constantly enhances workforce skills such that workers can fabricate complex, intricate, and unique prototypes and products using general-purpose equipment.

3.2.2 Batch Production

Batch production is the most common form of manufacturing systems. This mode of production is characterized by the fact that medium-size lots, hence the term *batch*, of the same component or product may be produced once, or they may be produced at some determined intervals. The manufacturing equipment used in batch production is general-purpose but capable of a higher rate of production. Since it is known ahead of time what products will be produced, the machine tools used in batch manufacture are often supported by specially designed jigs and fixtures, special tooling, and a good degree of automation or programmability. They greatly facilitate work, and the production rates can be noticeably high. A certain level of operator skill is also necessary because operators switch from product to product and deal with a variety of materials, equipment, tools, and manufacturing requirements.

The production process in this type of systems can be defined as the manufacturing of batches of identical parts by passing them through a series of operations such that, preferably, each operation is completed on an entire batch before it moves to its next operation. However, if deemed beneficial and appropriate, provisions are made so that sub-batches of parts can be transferred to their subsequent operation to improve the flow of work. Batch production is relevant because a large variety of parts and products will be needed in small quantities. The purpose of batch production is sometimes to satisfy the continuous or seasonally continuous demand for a product. Such a plant, however, is often capable of a production rate that exceeds the demand rate. Therefore, the plant first produces sufficient inventory of an item to satisfy the continuous demand; then it switches over to other products. This cycle is repeated, based on demand, the inventory levels desired, and the production rate. If the demand is seasonal, the production is accordingly coordinated. Examples of products made in a batch production environment include many types of industrial supplies, products with a different size and model, and household appliances, especially of seasonal type, such as heating units for winter and cooling units for summer. Batch production plants include machine shops, casting shops, plastic-molding factories producing containers for other industries, plants producing various colors of paint, and some chemical plants.

3.2.3 Mass Production

In mass production systems, manufacturing facilities are permanently allocated to produce a single part or product. Mass production is generally characterized by a large volume of output using specialized and often automatic equipment and machinery. This allows a low level of operator skill to be employed; the work involves very few tasks, often only loading/unloading of parts or supervising the operation of equipment. Commonly an entire plant is arranged with various production lines dedicated to producing different parts of one product to be assembled into a complete product. The investment in highly automated machines, integrated part transfer, and specialized tooling is significant. Compared with any other form of manufacturing, the skill level of workers in a mass production plant tends to be low. Because of specialized equipment and specific tooling for a particular product, flexibility to change to a different product, although possible, is low and could be costly.

Table 3.1 summarizes some of the important characteristics of job shop, batch, and mass production systems. The above description of these three types of production systems has been simplified. In practice, different approaches in the organization of production facilities must be used to manufacture goods as they rarely fall into one single category; that is, elements of one system might be present in the other. The reason is that it is not always possible, and in fact not necessary to maintain only one type of production system. For example, a dominantly batch manufacturing facility may have a mass or continuous production line for items that are used in large quantities within the plants. Similarly, a manufacturing facility may have two major sections, one section dedicated to the mass production of various products, and the other a batch manufacturing environment that can adapt to the demand variation to ensure that all products are available at all times.

Table 3.1. **Characteristics of different production systems**

Characteristic	Production System		
	Job Shop	**Batch**	**Mass**
Production quantity	Low	High	Higher
Production rate	Low	High	Higher
Workforce skill	Very high	High	Low
Equipment	General-purpose	General-purpose	Special-purpose
Tooling	General-purpose	Special-purpose	Special-purpose
Plant layout	Functional (process)	Functional or group technology	Line (product)
Degree of automation	Low	High	Higher

Figure 3.1 shows a typical range of application of machine tools and equipment within the conventional manufacturing systems in terms of production volume.

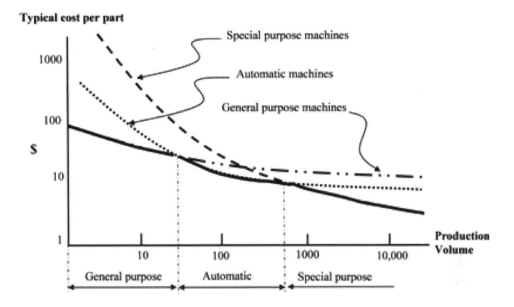

Figure 3.1. Application range of machine tools in production.

3.3 PHYSICAL ARRANGEMENT OF MANUFACTURING EQUIPMENT

The physical arrangement of the manufacturing equipment within a production facility is referred to as a "plant layout," and it is determined by its production system. A plant layout that is most suited to mass production would be quite impractical for job shop production and vice versa. There are four types of plant layout for the conventional manufacturing systems:

1. Fixed-position layout
2. Functional (process) layout
3. Group Technology layout
4. Line (product) layout

3.3.1 Fixed-Position Layout

In this form of layout, the product remains in a "fixed position" because of its large size and high weight. The equipment required for manufacturing and assembly is moved to and away from the product, as shown in Figure 3.2. Typical industrial scenarios include shipbuilding and aircraft manufacturing, which are characterized by a relatively low production volume and rate. In fixed-position layout, consumables such as nuts and bolts, tools, materials, prefabricated modules, and equipment must be arranged in a carefully organized fashion around the product while ensuring convenient access and open and safe pathways for unrestricted movement of materials, hardware, and personnel. As the production and assembly proceeds, congestion may ensue. Therefore, task planning and scheduling become increasingly critical.

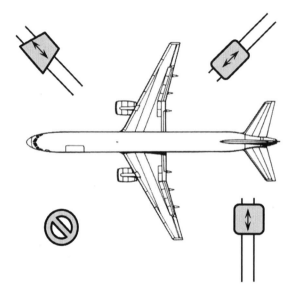

Figure 3.2. Fixed-position layout.

3.3.2 Functional Layout

The functional layout, which is primarily used for batch production, is the result of grouping production equipment or plant according to its function or type. Therefore, all milling machines, for example, would form one group, and all drilling machines would form another functional group, and so on. Figure 3.3 shows an example for this type of layout.

The functional layout improves work and operator control to obtain high efficiency in the utilization of the equipment and workforce in the plant. It also increases the flexibility of the plant owing to groups of highly skilled operators working with machines of similar functions in producing a wide variety of products.

Legend for Figures 3.3 to 3.8:

A   Assembly   B   Broach   D   Drill   G   Grinder   L   Lathe
H   Honer   I   Inspection M   Mill   T   Test/Tools S   Storage

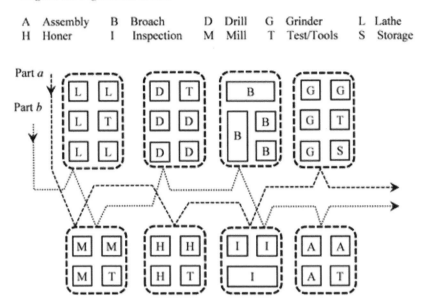

Figure 3.3. **Functional layout.**

In the functional layout, general-purpose rather than special-purpose machines are used. Thus, a relatively low capital investment is required. The nature of the functional layout permits the physical isolation of the unpleasant processes, such as some welding operations, and critical processes, such as explosive forming. This type of layout leads to the following characteristics:

- Need for highly skilled operators capable of performing a number of tasks on a variety of machines
- Many parts and products being produced at the same time
- Extensive storage space needed for raw material and semifinished parts
- Considerable storage and work space around machines
- High inventories of work-in-progress parts
- Need for various part-handling implements and transportation equipment
- Complex scheduling of work and tracking of materials flow through the plant
- Great flexibility in terms of producing different parts
- Frequent movement of materials between operations and departments, with the possibility of damage to parts and misplacement

3.3.3 Group Technology Layout

The group technology approach to manufacture products made in batches represents a major form of production comparable in importance to line layout for mass-produced goods. The key feature of group technology is group layout, in which the plant is arranged in groups of dissimilar types of machines according to complete manufacturing requirements of a group of similar parts known as component, production, or part family. The machine groups are frequently referred to as group technology cells. Figure 3.4 illustrates an example of group layout in component manufacture.

Group technology is a relatively new concept where families of parts that have some common characteristics are manufactured within designated cells. Figures 3.5 and 3.6 show two of these part families. A part family consists of a "group" of parts that have similar geometric shape and size features or have similar processing steps in their manufacturing. Parts shown in Figure 3.5 resemble a similar design family, but each could be different in their manufacturing requirements, such as material type and tolerances. In contrast, the parts shown in Figure 3.6 form a part family with similar manufacturing requirements.

Group technology has the advantage in terms of reduced material handling and provides a degree of specialization and centralization of responsibility for the complete manufacture of components. In group technology, manufacturing control within the cell is greatly simplified and work-in-progress and throughput times are reduced. However, group technology cells can be vulnerable in the case of an individual machine tool breakdown. Group technology cells should

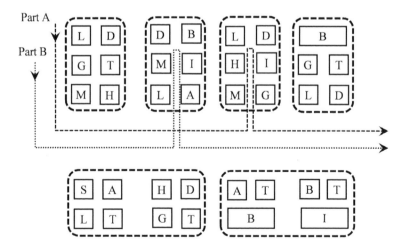

Figure 3.4. Group technology layout.

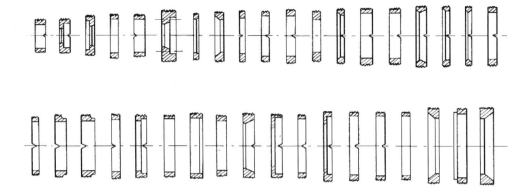

Figure 3.5. A design family.

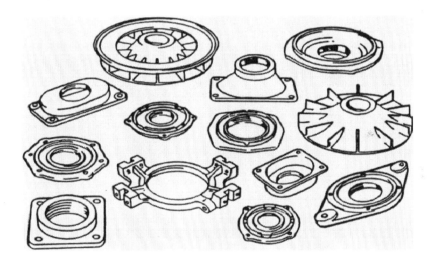

Figure 3.6. A production family.

be designed based on a careful analysis of total manufacturing requirements to identify component families and machine groups.

There are two other variations for cell organization. One is the "Single Machine Cell," which can be used for parts whose attributes allow them to be made on basically one piece of equipment, such as turning, grinding, or a multifunction machining center. The other form is the "Flow Line Cell," which is described in Section 3.3.4.

Many conventional manufacturing systems fall under the functional layout through evolution. When a bottleneck in the production capacity is observed, an additional machine is acquired and placed next to the existing machines of the same type. However, if this process continues, at some point the movement of personnel, materials, and parts becomes complex and inefficient. It would certainly be an enormous undertaking, but if a firm decides to reorganize its functional layout plant into a group technology layout, once the problems of part family identification and machine grouping are resolved, and the group technology concept and layout employed, a wide range of benefits are realized, including the following:

- Fewer design variations of parts produced
- Fewer variety of expendable items such as nuts, bolts, and washers
- More permanent or semipermanent tooling and machine setups
- Convenient and less congested material handling
- Better production and inventory control
- More efficient smaller process planning tasks
- Employee skill enhancement and job satisfaction

3.3.4 Line Layout and Group Technology Flow Line

With a line layout, the plant is divided into groups on a component basis similar to the case of the group layout. The order of processes to be performed on the component or product is the determining factor in the physical arrangement of manufacturing equipment. With conventional mass production, these groups of machines are arranged in lines, each permanently set up to produce one component. The high efficiency of mass production is generally the result of using a line layout where conveyors or gravity shoots are used to move parts from one machine to the next quickly. In a well-paced line, very high production rates are achieved, and the equipment and machines are rarely idle.

The line layout is also used for group technology families of similar components in such a way that every part in the family must (ideally) use the same machines in the same sequence. Certain steps can be omitted, but the flow of work through the system must be in one direction. This particular layout is frequently referred to as *group technology flow line*. Figure 3.7 shows an example of this type of layout.

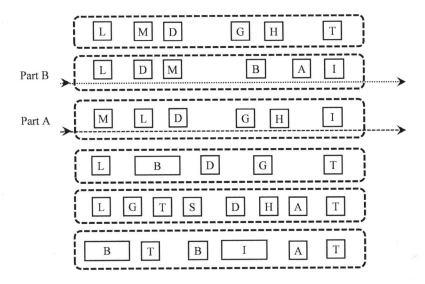

Part B

Part A

Figure 3.7. Group technology flowline layout.

3.3.5 Comparison of Plant Layouts

In the previous sections, individual characteristics of the three main manufacturing systems and their respective layout types were explained. Figures 3.8 and 3.9 and Table 3.2 aid in the comparison of these characteristics.

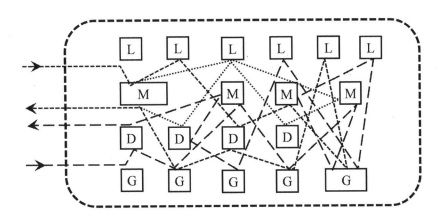

Figure 3.8. Complex material flow with a functional layout.

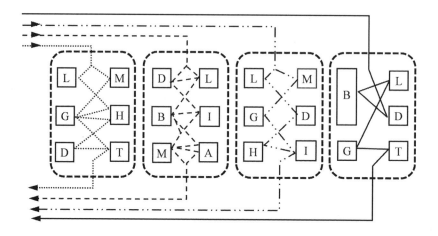

Figure 3.9. Simplified material flow with group layout.

Table 3.2. Characteristics of different layout types

Characteristic	Type of Layout		
	Functional	**Group**	**Line**
Specialization	by process	by component type	by component type
Material flow between machines	in batches	in batches to nearly continuous	continuous
Material throughput time	long	short	shorter
Stocks for work in progress	high	low	lower
Responsibility for quality	many persons per part	one person per part	one person per part
Responsibility for delivery by due date	many persons per part	one person per part	one person per part
Investment in special tooling	high: one set per operation/ part	low: one set per tooling family	high: one set per line; one line/part
Investment in buildings	high	low	lower
Control of material flow	complex	simple	simpler
Workforce skill	high	high	low
Job satisfaction	medium-high	high	low
Flexibility	high	medium-high	very low

3.3.6 Hybrid and Nested Manufacturing Systems

Purely functional or purely group technology layout may be rare in industrial settings. Commonly a production system is composed of elements of different manufacturing concepts and layout type, each complementing the other in an attempt to balance the particular situation's need with an economic solution. An expensive and less frequently used facility might be shared between a number of independent cells, or within a functional group, there may be machines with different functions. Certain components might only be produced by a certain subgroup of machines. On the other hand, within a group technology cell, a number of similar machines could be installed to balance the production capacity for particular components. The key is to pay attention to the pros and cons of any deviation from standard forms, and the ultimate goal should be to improve functionality and efficiency.

3.4 MODERN MANUFACTURING SYSTEMS

With the advent of computers, development of a multitude of sensors, and possibility of networking and real-time data transmission, manufacturing methods and systems have been tremendously revolutionized over the last few decades. Such manufacturing systems are commonly referred to as the Computer Integrated Manufacturing (CIM). In CIM, many facets of a manufacturing concern, such as process planning, inventory control, machining codes, in-process quality control, and some assembly, are integrated providing a high degree of efficiency, coordination, and productivity. Within CIM there exists the fully automated category of Flexible Manufacturing Systems (FMSs).

3.5 FLEXIBLE MANUFACTURING SYSTEMS

FMSs come in various designs, sizes, combinations of equipment, tooling provisions, material storage and handling, and specific unique features. The key characteristics of FMSs are the following:

1. Group technology, where a subtle variety of parts can be manufactured. This is because the system is primarily and often permanently fitted with tools and auxiliary devices that handle a known range of weight, volume, and shape of components.

2. Computer-controlled material-handling equipment and/or robots. These implements are generally immediately available to load, unload, or transfer parts throughout the system.

3. Computer-controlled workstation of various types. The cycle time per part on the workstations must be highly synchronized and coordinated so that they don't stay idle for too long or block each other as a result of having accumulated unfinished parts.

4. Loading and unloading stations. The stations that in some systems are used as temporary buffers are integrated with the material-handling system for a continuous, smooth operation.

5. Coordinated operation. All the above elements must be highly synchronized, have compatible cycle times, well integrated, and have "look ahead" intelligence to prevent blockage, and, if an unpredictable blockage occurs, have means of unblocking and resuming operation.

Despite the term *flexible* (and indeed these systems are highly flexible), their flexibility is within a narrow range of the parts that they can produce. With constant developments in technology, their flexibility has been broadened, and a greater variety of parts can be produced within these systems.

3.5.1 Advantages of FMSs

With known-in-advance parts to produce and highly coordinated operations, it is possible to gain major advantages using FMSs, such as the following:

- Low manufacturing cost per part due to dedicated tooling and machinery

- High machine utilization due to synchronized cycle times

- Improved and more consistent quality due to absence of direct human involvement

- Increased system reliability due to self-adaptation, quick replanning and resumption of operation, and elimination of human errors

- Reduced part inventories and work in progress due to automated handling and synchronized operations

- Short work flow lead times due to elimination of idle time and continuous activities in the system

- Significantly improved overall efficiency considering the above factors

Figure 3.10 shows the percentage of time spent by a typical part in conventional manufacturing facilities from entry to exit. Effectively, it is 30 percent of 5 percent, that is, 1.5 percent of net machining or processing time, leading to significant costs and minimal returns. In well-synchronized flexible manufacturing, the net machining and processing times could be as high as 80 percent.

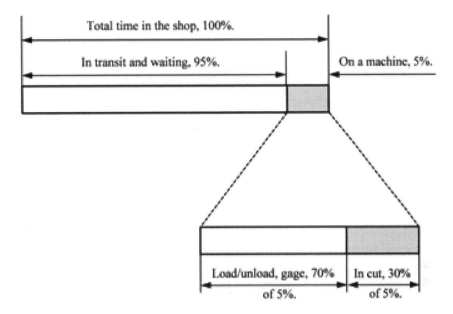

Figure 3.10. Breakdown of time spent by a representative part in various stages in a conventional manufacturing facilities.

3.5.2 Disadvantages of FMSs

FMSs are not without some disadvantages despite of the significant benefits they offer. Notably, FMSs entail:

- A high capital cost of implementation, equipment, and infrastructure
- Substantial programming, planning, and problem-avoidance efforts
- Few but very skilled operators to maintain operations
- A complex system of interconnections requiring significant service, repair, and maintenance efforts

3.6 PHYSICAL CONFIGURATION OF FLEXIBLE MANUFACTURING SYSTEMS

The physical arrangement or layouts for FMSs can be diverse, and innovative configurations are possible. The key factor in any particular design is to plan for a congestion-free material handling that interconnects workstations and load/unload areas.

3.6.1 Cell Layout

The simplest form of FMSs is the flexible manufacturing cell, where almost all the key material handling is performed by one or very few robots. The most common is a single-robot cell, where the robot is located within a group of workstations serving all as depicted in Figure 3.11. Some may be supported by an input-output table, a conveyor, or a carousel for parts.

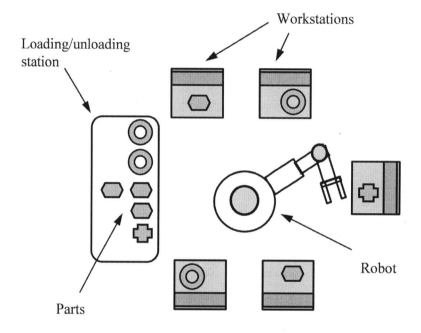

Figure 3.11. **A flexible manufacturing cell.**

3.6.2 Linear Layout

The term *linear* refers to the straight line track along which material handling transporters move back and forth. In a linear layout, the machine tools, workstation, and auxiliary equipment are located on either one or both sides of the material-handling line. Roller-type conveyors, or tracked transporters, as shown in Figure 3.12, are common.

3.6.3 Loop Layout

With the loop layout, material handling is through a closed loop system where raw material, semifinished, and finished parts generally circulate in one direction, as

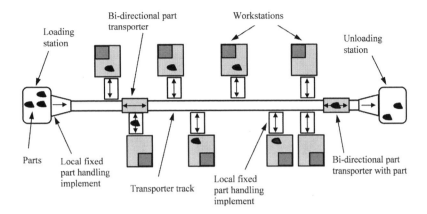

Figure 3.12. **Linear layout FMS.**

shown in Figure 3.13, though bidirectional circulation is also possible. If the loop system has cross connections that result in smaller multiple loops sharing paths, because of its shape, it is referred to as the "ladder"-type layout.

3.6.4 Carousel System

In this configuration, a centrally located multilevel carousel is used. By revolving as needed, the carousel accepts, temporarily stores, and delivers parts to the

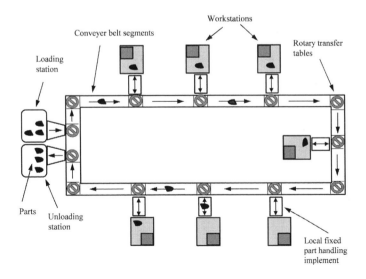

Figure 3.13. **Loop layout FMS.**

machining facilities. In this innovative design, multiple robots are used for all loading and unloading of parts from the storage and workstations around the carousel. Figure 3.14 shows a schematic of this form of FMS.

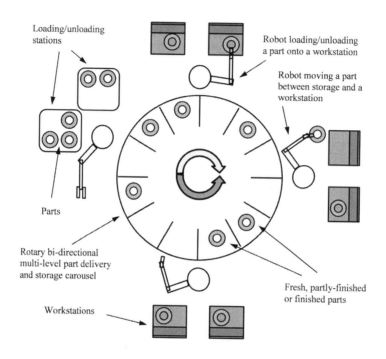

Figure 3.14. Carousel type FMS.

3.6.5 Other Layouts and Systems

There is no limit to innovative designs for FMSs in capability and layout type. In these systems, however, for continuous machining operations the chip removal from the cutting zone is perhaps one of the most challenging tasks. With the use of chip-beaker tools and tool bits, this can be resolved to some extent. Nevertheless, chips will still be produced at a high rate and often accumulated on the machining table. Use of coolant fluid and blast of air can displace most of the chips. However, any chip-pieces remaining on the machining bed can cause improper placement of workpieces on the machine. One idea to reduce the possibility of this occurring is to use workstations that are either installed entirely upside down or have a cutting module that can be rotated 180 degrees so that all chips fall off into the ground collection basins. Additionally, cutting coolant can assist the process further by washing away the chips produced.

This form of FMS requires specially built machinery and workstations that are expected to have significant cost implications.

3.7 General Comments

The typical systems shown were simplified to demonstrate some of the general configurations possible. In practice, many support and auxiliary units, not explicitly shown in the figures, are necessary to facilitate the functioning of an FMS. These include the following:

- A central computer that coordinates and controls the entire operation
- Palletizing/de-palletizing of parts
- A tool crib for workstations
- Intermediate wash and degreasing stations
- Means for effective problem-free chip removal in machining operations

EXERCISES

3-1 In your place of study or work, a university machine shop, a small or a large manufacturing plant, determine the form of the manufacturing system that is in place and whether its arrangement of equipment is closest to the line, functional, or group technology layout.

3-2 In a follow-up to Exercise 3-1, investigate whether another type of layout would be a more suitable choice.

3-3 In which type of manufacturing systems is scheduling of work simpler?

3-4 Comparing functional and group technology layouts, which one offers better opportunity for control and responsibility over the parts or products?

3-5 If your place of work is a conventional manufacturing environment, and you were to propose a plan to convert it to an FMS, which layout form will you choose and why?

3-6 What does the statement "future products will be mass-produced in batches of one" mean?

3-7 Why does a typical part spend only 1.5 percent of its time in a conventional manufacturing system in the actual fabrication process?

Chapter 4

CLASSIFICATION AND CODING

4.1 INTRODUCTION

Classification and coding method, or, more specifically, part classification and coding method, entails identifying the similarities among a large number of parts and manufactured pieces, and expressing these similarities through a structured coding system. In production plants, two types of part similarities are of interest: similarities that relate to the design features, such as geometric shape and size, and similarities that stem from manufacturing attributes and relate to the type and sequence of production processes required to fabricate the part. In addition to design and manufacturing classification, which already include some material properties, dedicated classification and coding systems can be developed and utilized for solid materials, chemical compounds and liquids, and to many other industrial materials and settings. For example, classification of chemical compounds can help in proper storage assignment to reduce the risk of reaction or help extinguish fires with appropriate retardants.

In manufacturing, using a classification and coding system can significantly reduce the number of different parts and tools required, simplify storage and retrieval, standardize design, and facilitate production and assembly. The extended benefits will be reviewed in detail later in this chapter.

One of the important applications of classification and coding is for group technology (see Chapter 3), where some classification effort is required to implement the group technology concept. We focus on part classification and coding using group technology as a prime example, but the concepts are general and can be used in a wide range of situations. With the use of common software such as Excel, a customized classification and coding system can be tailored for modest cases and developed in-house.

4.2 GROUP TECHNOLOGY

We saw in Chapter 3 that group technology is a concept in which similar parts are grouped together to benefit from their similarities in design and manufacturing. For instance, in a plant producing five thousand different parts, it may be possible to group a vast majority of these parts into, say, fifty distinct part families. Some parts may not readily fit into a particular family, and the production personnel must carefully assess these for special accommodation. For the part families formed, the required production equipment is rearranged into groups known as cells. Each cell possesses a variety of machines necessary to process one or several families of parts and functions fairly independently from the other cells. Normally, families of parts are released to the cell one family at a time and in predetermined batch sizes.

4.2.1 Part Families

We defined a part family as a group of parts that are similar in geometric shape, form, and size, or because similar processing steps are required in their manufacture. Most existing or traditional plants are laid out in the functional form. In order to implement the group technology concept, a changeover is necessary. The classification and coding method is the most efficient way to accomplish this and generate appropriate design and production part families. The method involves classifying a population of parts into families by examining the individual design and/or manufacturing attributes, or CAD-CAM records of each part. The classification results in a series of digits or a code number that uniquely identifies the part's attributes. A number of different classification and coding schemes have been devised, and there are also several commercially available systems. In the following sections, we review key points and benefits of implementing the concept of classification and coding and outline its various coding structures.

4.2.2 Part Classification and Coding

Implementing a classification and coding system requires effort and commitment of time regardless of the method used. From the general concepts to specially developed software, none has become an industry standard because the requirements of various companies differ in nature. However, application of any classification and coding system in group technology provides a range of benefits, including the following:

1. It facilitates the generation of part families and machine groups or cells.

2. It permits quick access and retrieval of CAD and CAM information.

3. It reduces design duplication and thus the variety of parts.

4. It provides reliable workpiece statistics.

5. It facilitates estimation of machine tool requirements and efficient utilization of the machines.

6. It permits streamlining of tooling, jigs, and fixtures, thus reducing setup and production throughput times.

7. It allows standardization and improvements in tool design.

8. It aids production planning and scheduling procedures.

9. It provides better machine tool utilization and better use of tools, fixtures, and manpower.

10. It facilitates part programming for numerically controlled (NC) machining.

The design and manufacturing attributes that are generally used in classification schemes for manufacturing are listed in Table 4.1. It is important to pay attention to the overlap between the design and manufacturing attributes. On one hand, it may not interfere with the classification effort at hand, and even provide some flexibility in choosing a part family. On the other hand, it can lead to a wrong classification decision being made.

Table 4.1. Attributes of two forms of part families

Part Design Attributes	Part Manufacturing Attributes
Basic external shape	Major processes
Basic internal shape	Minor processes
Length/diameter ratio	Major dimension
Features for assembly	Length/diameter ratio
Features for functionality	Surface finish
Material type	Machine tools required
Part function	Operation sequence
Major dimensions	Production time
Minor dimensions	Batch size
Tolerances	Production volume
Surface finish	Fixtures required
	Cutting tools required

4.3 CLASSIFICATION AND CODING SCHEMES

A number of classification and coding schemes have been developed that can be categorized into the following forms:

- Monocodes
- Polycodes
- Hybrids

4.3.1 Monocodes

In the monocode scheme, each digit of the code successively represents a feature and its subgroups. Monocode systems are also referred to as hierarchical because the expansion of the codes results in a treelike structure. Each digit or node in the code depends on the previous node. Monocodes are efficient is subdividing a feature into its constituent groups, but pose a difficulty in interpreting individual nodes due to their dependence on the previous node. The basic structure of the monocode scheme is shown in Figure 4.1.

Advantages:

- Using only a few digits a large amount of information can be recorded.
- Various parts (or branches in the hierarchical structure) of the code can be utilized to extract a particular subset of the information stored.

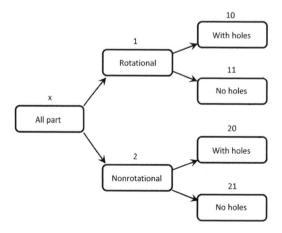

Figure 4.1. **Structure of monocode classification and coding scheme.**

Disadvantages:

- It poses difficulty in consistently expanding the hierarchical structure.
- If a feature or subgroup does not expand further, bank codes or dummy codes must be used.

4.3.2 Polycodes:

This scheme is structured on an entirely different concept than monocodes. In the polycode scheme each digit or node represents one independent feature of the part. Figure 4.2 shows an example of how each digit has a description that is unrelated to the description of the other digits.

Advantages:

- The greatest feature of this scheme is its ease of devising and using.
- It can be readily developed for applications of modest size.

Disadvantages:

- Unlike in monocodes, where each digit inherits information from the preceding digits and carries more information, to achieve the same level of information in polycodes, a much longer code string is necessary.

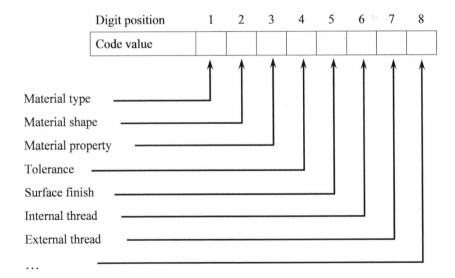

Figure 4.2. Structure of polycode classification and coding scheme.

- Comparison of segments of code to extract similarity information between parts requires effort. In particular, if the code is custom-engineered and long, a computer program may be needed.

4.3.3 Hybrids:

This structure does not have the disadvantages of the monocode and polycode concepts. However, users ought to examine all three schemes and determine which one best satisfies their classification and coding needs. An example of a hybrid classification and coding structure is shown in Figure 4.3.

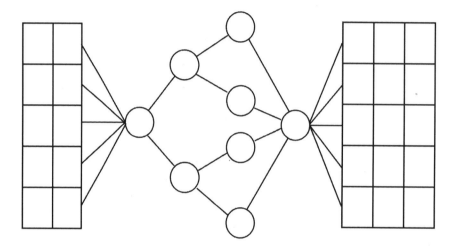

Figure 4.3. **Generic structure of hybrid classification and coding scheme.**

Advantages:

- Hybrid structures combine the benefits of monocode and polycode systems, thereby providing flexibility to fit a variety of applications.
- Most commercial systems are of the hybrid type, hence providing greater choice.

Disadvantage:

- There are no specific disadvantages.

4.3.4 The Opitz Classification System

Developed by H. Opitz of the University of Aachen in Germany, and named after him, the Opitz classification and coding system is considered one of the pioneering efforts in the field. It has become a popular platform and a generic model in introducing the fundamental concepts of classification and coding. The Opitz coding structure is of hybrid type and consists of the following three-part digit sequence:

$$12345\ 6789\ ABCD$$

The first group of five digits, 12345, is referred to as the "form code" and describes the geometry and design attributes of the part. The next group of digits, 6789, represents the "supplementary code," which specifies some of the attributes that must be known to manufacture the part. The last group of digits, ABCD, is designated as the "secondary code" and is used to indicate the production processes and their sequence. The secondary code can be tailored to serve specific needs of a company. The interpretation of the first nine digits is given in Figure 4.4.

The code definition for the first five digits for the rotational part classes 0, 1 and 2 (digit 1 in Figure 4.4) is given in Figure 4.5. The complete coding system covering all part classes and their code description is extensive.

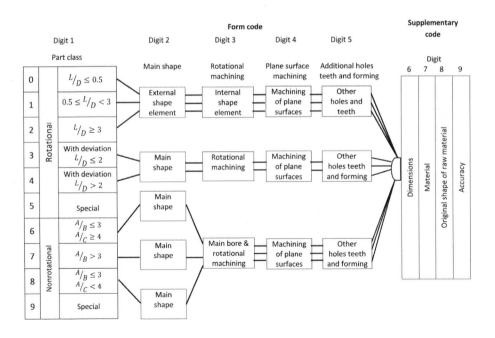

Figure 4.4. **The core structure of the Opitz part classification and coding system.**

	Digit 1 Part class		Digit 2 External shape, external shape elements		Digit 3 Internal shape, internal shape elements		Digit 4 Plane surface machining		Digit 5 Auxiliary holes and gear teeth
0	$L/D \leq 0.5$	0	Smooth, no shape elements	0	No hole, no break through	0	No surface machining	0	No auxiliary hole
1	$0.5 < L/D < 3$	1	No shape elements	1	No shape elements	1	Surface plane and/or curved in one direction, external	1	Axial, not on pitch circle diameter
2	$L/D \geq 3$	2	Thread	2	Thread	2	External plane surface related by graduation around a circle	2	Axial on pitch circle diameter
3		3	Functional groove	3	Functional groove	3	External groove and/or slot	3	Radial, not on pitch circle diameter
4		4	No shape elements	4	No shape elements	4	External spline (polygon)	4	Axial and/or radial and/or other direction
5		5	Thread	5	Thread	5	External plane surface and/or slot, external spline	5	Axial and/or radial on PCD and/or other direction
6		6	Functional groove	6	Functional groove	6	Internal plane surface and/or slot	6	Spur gear teeth
7		7	Functional cone	7	Functional cone	7	Internal spline (polygon)	7	Bevel gear teeth
8		8	Operating thread	8	Operating thread	8	Internal and external polygon, groove and/or slot	8	Other gear teeth
9		9	All others	9	All others	9	All others	9	All others

Figure 4.5. **Form code definitions for part classes 0, 1, and 2 in the Opitz system.**

Referring concurrently to Figures 4.4 and 4.5, and starting from digit 1, we can see how the coding system progressively adds information to the resulting part code. For instance, digit 1 identifies whether the part is rotational or nonrotational (prismatic). By using the ratios of the part's major dimensions, it also indicates whether the part is slender or not. After digit 1 has been assigned, digit 2 is selected in relation to it, and so on in a successive manner. Example 4.1 demonstrates the application of the Opitz coding system using the definitions given in Figures 4.4 and 4.5.

Example 4.1

Determine the "form code" for the workpiece shown in Figure 4.6 using the Opitz coding system.

Solution

The envelope length/diameter or aspect ratio, L/D = 1.25, suggests the first digit code as 1. The part is stepped on one end with no shape elements, so the second

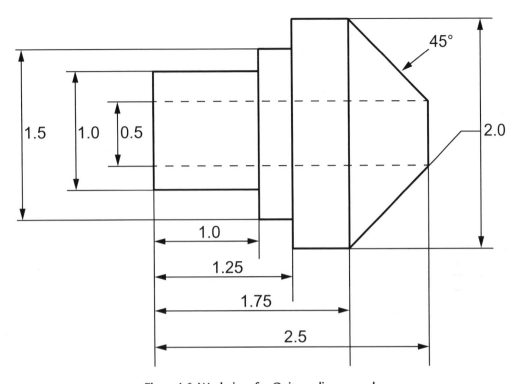

Figure 4.6. Workpiece for Opitz coding example.

digit code would be 1. The third digit code is also 1 because of the smooth through-hole. The fourth digit code is 0 since no surface machining is required, and the fifth digit code is also 0 as the part has no auxiliary holes or gear teeth. The form code in the Opitz system for this part is 11100.

4.3.5 Commercially Developed Systems

In this section, we briefly introduce three commercially developed classification and coding systems.

MICLASS

MICLASS stands for Metal Institute Classification System and was developed in the Netherlands, and maintained and supported by the Organization for Industrial Research. Its purpose was to help in automating and standardizing design, production, and management. MICLASS is composed of thirty digits. The first four digits have a monocode form, and the rest of the digits have a polycode structure. The first twelve digits have been designated for predefined purposes. The remaining eighteen digits provide great flexibility to store company-specific data. The coding construct of the first twelve digits is shown in Figure 4.7.

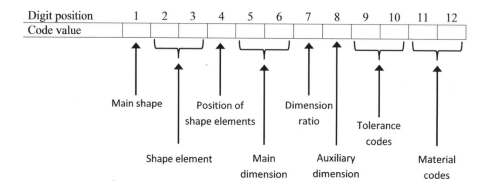

Figure 4.7. The structure for the first twelve digits of the MICLASS system.

KK3

This classification and coding system was developed by the Japan Society for the Promotion of Industrial Machines. The primary application of KK3 is the general classification of machined metal parts. It uses a twenty-one-digit hybrid code structure shown in Figure 4.8.

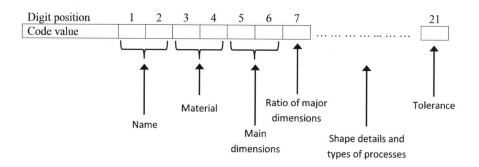

Figure 4.8. The code structure of the KK3 system.

MultiClass

MultiClass is a classification and coding system developed by the Organization for Industrial Research. It is the newest and most flexible of the three systems and allows the users to customize the coding system to fit their requirements and applications. MultiClass can be used for applications in manufacturing machined and sheet metal parts, tooling, electronics, and assemblies.

MultiClass uses a monocode or hierarchical coding structure and has a menu-driven software. The same as MICLASS, its structure consists of up to thirty digits. The first eighteen digits have a preassigned use (see Figure 4.9), and the remaining twelve digits can be defined by the user to code additional information. A prefix (digit 0) precedes the thirty-digit string to categorize parts into major classes.

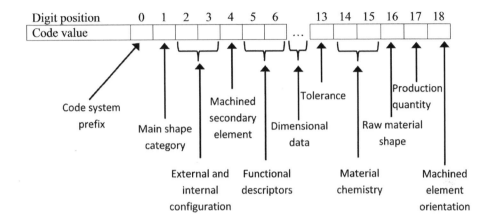

Figure 4.9. **The structure for the first eighteen digits of the MultiClass system.**

4.3.6 Custom-Engineered Classification and Coding Systems

For a modest-size application, if a commercial system is not available or deemed unsuitable, an enthusiastic and determined industrial engineer can custom-design their own classification scheme. Using software such as Excel, a practical classification and coding system can be developed in-house. Since the monocode scheme stores more information and quickly categorizes the part population into major classes than the polycode scheme, in custom-designing a classification and coding system it may be worthwhile to giving consideration, at least for the first few digits, to have a monocode structure.

EXERCISES

4-1   To assist in convenient and safe storage of an inventory of a large number of different chemicals, develop a five-digit Opitz-like code with the following features:

    i. Select five characteristics that you think are the most important to consider. Say, for example, flammability, volatility, and reactivity.

    ii. Assign each of these characteristics to one of the digits of the code. For example, digit 2 (from left) could be flammability, volatility, and reactivity.

    iii. For each digit, then assign at least three classes. For instance, for the flammability:

        0   = Nonflammable

        1   = Semiflammable (that is, flammable under certain conditions, such as being exposed to open flames for an extended period of time)

        2   = Flammable

4-2   For the machined part shown in Figure 4.10, define the five-digit "form code" using the Opitz system.

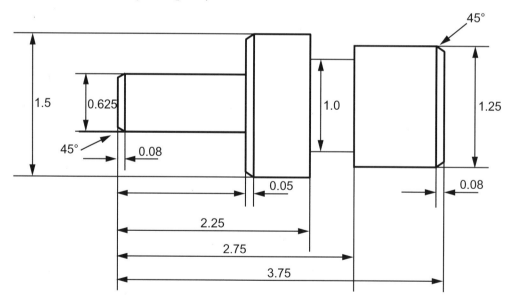

Figure 4.10. Part drawing for Exercise 4-2.

4-3   Select an application area of your interest or from your experience, and design a coding system with a relevant classification similar to Exercise 4-1.

Chapter 5
SEQUENCING AND SCHEDULING OF OPERATIONS

5.1 INTRODUCTION

The words *sequencing* and *scheduling* are often used interchangeably to the point that sometimes they seem indistinguishable. But they are, in fact, two different concepts. *Sequencing* refers to the order determined to carry out a set of operations or tasks, whereas *scheduling*, relative to a reference time, indicates the start, duration, and the end time for each task. We also use these terms having the same meaning in our explanations, but when drawing a time chart, the difference will become apparent.

To appreciate the importance of scheduling, it is perhaps more effective to think of the consequences of the lack of it. For instance, in a manufacturing facility, the resources will be underutilized, delivery promises will be missed, and client dissatisfaction will ensue.

Some concepts of sequencing and scheduling are straightforward and practical, while others are complex, mathematically involved, and mostly of academic nature.

Scheduling or sequencing is an important aspect of production control and providing services, in determining the order and the correct use of facilities within each instant of time. The primary goal is to make the greatest use of the facilities available by maintaining a balanced and concurrent level of activities in all machines, equipment, and departments so that expensive resources and skills are seldom idle. However, the specific objectives in scheduling vary from company to company and from case to case.

Many factors have to be taken into consideration when making decisions to achieve certain scheduling objectives, such as availability of resources, cost, and implications of implementing a decision. What makes scheduling such an essential function to improve overall utilization is that resources are not always available. A plant or a machine can be in different states over a selected life cycle. A twenty-four-hour examination of these states for a machine tool is depicted in Figure 5.1.

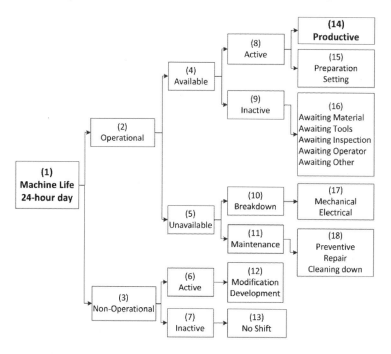

Figure 5.1. Functional breakdown of a twenty-four-hour life-cycle of a machine tool.

It is clear that more than twenty different conditions can be identified, out of which only one state is "productive." Statistical and historical data show that for conventional machine tools this is in the order of 1.5 percent (See Section 3.5.1 and Figure 3.10). Even if one contests the statistics and assumes this ten times larger at 15 percent, the 85 percent lost opportunity is still substantial. Therefore, it is important that the valuable productive time is synchronized with the productive time of other resources through scheduling to improve the overall efficiency and productivity.

A considerable amount of research has been conducted in the area of scheduling, or, as it is more often referred to, "job shop scheduling and sequencing of operations." No single-most effective solution can be developed because of the diversity of problems. The scale of the problems rises further, and complexity is added when practical aspects of production are also considered. To counter this to some extent and make the development of a schedule feasible, many simplifying assumptions are often made. One must, however, pay attention to these simplifications because too many assumptions can diminish the relevance and value of the determined schedule. Figure 5.2 shows a view of the scope of scheduling solutions.

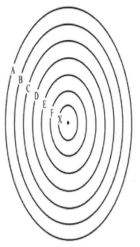

A ⇒ Infinite number of possible schedules

B ⇒ Reduced number of schedules by stating objective function(s)

C ⇒ Reduced number of schedules by making simplifying assumptions

D ⇒ Separate smaller schedules by breaking down work into groups

E ⇒ Reduced number of schedules by using algorithms

F ⇒ Reduced number of schedules by using heuristic rules

X ⇒ Achieved schedules with manipulation

• ⇒ Target

Figure 5.2. **The extent of scheduling solutions.**

Given the complexity of practical scheduling problems, the stochastic nature of nonproductive states shown in Figure 5.1 can lead to the deterioration of an existing schedule. Typically, generating a new schedule or modifying an existing one, among other things, can be an inconvenient task in the least, and the constantly changing priorities and constraints of the scheduling problem make the prospect of a practical and an optimal dynamically responsive schedule an elusive ideal.

Despite these issues, working without a schedule that can shed light on the state of affairs is a hazardous choice. It is interesting to note that efficient and intelligent solutions for thousands of scenarios have been developed.

Some tend to be theoretical, but many have practical applications. In this chapter, we review several important categories of scheduling and sequencing problems, their classification, and solution algorithms. Many algorithms exist for various situations, but they often have a number of simplifying assumptions. Despite such limitations, the analysis and application of such methods will help us understand, apply some of the ideas in practice, and, especially when some experience has been gained, even develop solutions with intuition and without resorting to any mathematical calculation.

5.2 DEFINITION OF SCHEDULING TERMS

To be able to formulate, address, classify, and develop solutions for scheduling problems, it is necessary to define and understand some standard terms and notations. The common terms and their definition are as follows.

- *Processing time.* The time required to complete a task on a machine or a process.

- *Due date.* The agreed delivery time or deadline for a task beyond which it would be considered late or tardy (see the definition below). Sometimes some form of penalty for being tardy may be applied.

- *Lateness.* The difference between a task's completion time and its due date will have a positive lateness if the task is completed after its due date and negative lateness if completed before its due date.

- *Tardiness.* This is the measure of positive lateness. If a task is early, it has negative lateness but zero tardiness. If a task has positive lateness, it has equal positive tardiness.

- *Slack.* A measure of the difference between the remaining time to a task's due date and its processing time.

- *Completion time.* The span between the beginning of work on the first job on the first machine, which is referred to as $t = 0$, and the time when the last task is finished on the last machine.

- *Flow time.* The time between the point at which a job is available for processing and the point at which it is completed. When all the jobs are available at time zero (also known as the Ready Time $r = 0$), then flow time is equal to the completion time.

In dealing with scheduling problems, we often work with these time-based terms. Figure 5.3 schematically shows the relationship between some of these terms.

From Figure 5.3 many relations can be written and used depending on the objectives of the problem at hand.

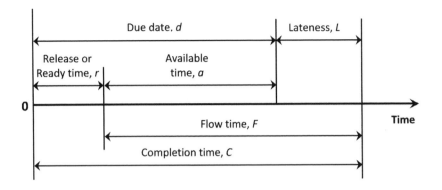

Figure 5.3. **Time-based terms in scheduling problems.**

$$a = D - r$$
$$F = C - r$$
$$L = C - d$$
$$L = F - a = C - r - a = C - (r + a) = C - d$$

When there is more than one job, these relations also stand in terms of averages for the group because all the relations are valid for any job. Therefore, each equation can be written n times (for the n jobs), summed, and averaged. For example, if we show the average of a parameter with a bar over the top the last relation can be written as:

$$\overline{L} = \overline{F} - \overline{a} = \overline{C} - \overline{r} - \overline{a} = \overline{C} - (\overline{r} + \overline{a}) = \overline{C} - \overline{d}$$

If lateness, L, is calculated as negative, it is referred to as earliness (a positive value E). Lateness is an important measure because penalties or rewards can be associated with it. If in specific situations earliness is not rewarded, a different measure is used for ease of mathematical manipulation, which is referred to as tardiness, T, and defined as:

$$T = Max \ (0, L) \geq 0$$

If the "ready time" is not specified, it is assumed as zero, in which case the flow time and completion time will be the same. The most frequently used measure of optimality for multimachine, multijob problems is the makespan, throughput, or completion time, C_{max}, which is the total amount of time required to completely process the entire set of jobs. The best schedule is the one that has the smallest C_{max} among the alternative schedules. Other common measures include minimizing the maximum lateness or minimizing the mean flow time.

In fact, for multimachine, multijob problems, it is often desired to achieve the minimum of the maximum or average of a measure, except earliness, where a maximum will generally be desirable.

It is also evident from the above set of equations that $\bar{L}$, $\bar{C}$, and $\bar{F}$ are equivalent. That is, if one is minimized, the other two will be, too. A performance measure, objective or criterion is always associated with any scheduling problem. Whatever the process is, a mechanical machining, a welding operation, a chemical reaction, or an assembly task, it is customary to use the term *machine* rather than *process*, and show their number by *m*. The various items, materials or entities of any kind that are processed on the machines are commonly referred to as the "jobs," and their number is shown by *n*. Also, the plant environment is called a "job shop." We will adhere to these conventions. A job shop consists of a set of general purpose machines that perform operations on various production orders or jobs. In this context we do not consider the operations sequence in a mass (continuous) production line, where an order for the machines is determined on a permanent basis, and job scheduling is immaterial as it will not affect the long-term production rates. Many of the job shop problems and the algorithms and solutions deal with a pool of *n* jobs, which are often all available at the same time, and a group of *m* machines.

5.3 SCHEDULING ALGORITHMS

Algorithms are a set of rational rules or iterative procedures applied to the data available to solve the related problem in a finite number of steps. Different algorithms can be developed for a specific scheduling problem leading to the same optimal or near-optimal solutions. Before introducing such algorithms for scheduling, let us examine what the alternative might be. A common criterion for a job shop is to minimize the completion time. Suppose it is desired to process three jobs, *a*, *b*, and *c*, through two machines, M_1 and M_2, in the shortest possible time with the restriction that for any job sequence the machine sequence will be fixed, that is, either M_1 is always first or M_2. The brute force approach is to generate all possible combinations. The combinations when one fixed sequence is used for both machines are shown in Figure 5.4. If the job sequence varies for the second machine, the total number of combinations will be seventy-two.

For each of the sequences a process time chart, commonly known as the Gantt chart, can be drawn and the smallest completion time selected as the optimal solution. This can also be done mathematically through a computer program. A sample Gantt chart for Figure 5.4 is shown in Figure 5.5.

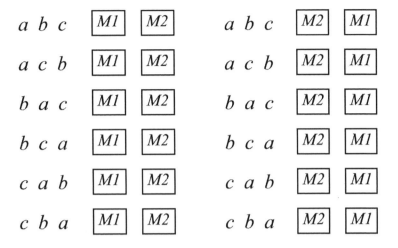

Figure 5.4. Possible fixed processing sequences for a three-job, two-machine problem.

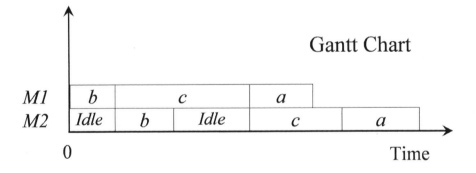

Figure 5.5. Sample Gantt chart for a three-job, two-machine problem.

5.3.1 Objectives in Scheduling Problems

The objectives in scheduling vary from case to case and often from day to day. Examples include the following:

- Achieving contractual commitments and meeting delivery promises or due dates
- Completing all tasks in the shortest possible time
- Finishing expensive and bulky jobs as early as possible or finishing tasks with other "rankable" attributes in a specified optimal fashion

- Maintaining a balanced level of activity in all machines/departments, so that all resources are utilized to the highest degree possible

5.3.2 Industrial Problems

Examples of industrial problems include the following:

- Scheduling jobs through manufacturing processes in a plant
- Executing different programs on a computer (time-sharing, file transfer traffic)
- Determining landing order of airplanes (based on fuel left, size of the aircraft, next departure time, type of runway available)
- Scheduling surgeries in operating rooms (shortest possible time, giving priority to the most ill)

5.3.3 The Practice

In design, analysis, and use of algorithms, simplifying assumptions are often made. For example, processing times are taken to be sequence independent. In practice, however, machine setup times can change depending on the sequence used. For instance, if jobs *a*, *b*, and *c* follow one another, and they are all made of the same material, then the "machine cleaning time" may be eliminated in between their processing—meaning that the actual processing time could be shorter than initially assumed. It is the industrial engineer's responsibility to note, discover, and use such time-saving factors involved in the real world.

5.4 *n*-JOB ONE-MACHINE PROBLEM

With one machine, it may appear that there should be no issues with scheduling the work through as the jobs can be processed in succession without any delay. But this depends on the measure and criterion of optimality. As far as the maximum completion time is concerned, any sequence will have the same result. However, with a measure such as minimizing mean lateness or mean flow time, the result will be different, and a specific sequence will be optimal. As an example, suppose we wish to minimize the mean flow time of processing *n* jobs through one machine.

Let:

P_i be the processing time for job i

$P_{[j]}$ be the processing time for a job placed in a position j

$F_{[k]}$ be the flow time of a job in k^{th} position of an arbitrary sequence

We show that ordering the jobs in a Shortest Processing Time First rule (SPTF, or simply SPT) sequence minimizes the mean flow time of the n jobs.

For a job in position k, the flow time will be the sum of processing time of the preceding jobs plus its own processing time, that is:

$$F_{[k]} = \sum_{j=1}^{k} P_{[j]}$$

The mean flow time for the group of n job in the sequence is:

$$\bar{F} = \frac{\sum_{k=1}^{n} F_{[k]}}{n} = \frac{\sum_{k=1}^{n} \sum_{j=1}^{k} P_{[j]}}{n}$$

Expansion and examination of the individual summation terms reveal that:

$P_{[1]}$ appears n times

$P_{[2]}$ appears $n-1$ times

...

$P_{[n]}$ appears 1 time

Then the above equation can be written as:

$$\bar{F} = \frac{\sum_{j=1}^{n} (n-j+1) P_{[j]}}{n}$$

From calculus, the sum of the pairwise product of two sequences of numbers is minimized by arranging one sequence in nonincreasing (deceasing or equal) order and the other in nondecreasing (increasing or equal) order. With i increasing from 1 to n, the multiplier $(n-i+1)$ results in a nonincreasing order; therefore, by sequencing the terms $P_{[j]}$ in nondecreasing order, the mean flow time $\bar{F}$ is minimized. In other words, we sort the original P_i, $i=1, n$ into a nondecreasing order as $P_{[j]}, j=1, n$. This is what we earlier referred to as the SPT rule. With a similar argument, we can also show that SPT minimizes mean lateness and mean waiting time.

5.4.1 Classification of Scheduling Problems

Now that we are familiar with the scheduling problem, we introduce a classification format that is used to express and identify various scheduling problems in the form of $n/m/A/B$ code, often used as a means of addressing scheduling problems and solutions in the literature among researchers and practitioners. In this coding system, the following is true:

- n is the number of jobs or tasks.
- m is the number of machines or processes.
- A indicates the job flow pattern through the machines.
- B signifies the performance measure and the criterion for its optimality.

There are three common forms of flow patterns, and the symbols for placement in the position A of the code are the following:

- F stands for the flow shop, where the machine order is the same for all the jobs.
- P stands for the permutation flow shop, where not only is the machine order the same for all the jobs but the jobs also have a fixed sequence throughout.
- G stands for general job shop, where there is no order restriction for the machines, or for the jobs.

When $m = 1$, A can be left blank. For the position B in the code, there are many options, such as C_{max}, F_{max}, L_{max}, $\overline{C}$, $\overline{F}$, $\overline{L}$. One must be careful in interpreting all performance measure symbols. For instance, C_{max} means *"minimize the maximum completion time"*. Many such terms are self-evident in meaning. Also, for average measures such as $\overline{L}$, it is customary to skip the subscribe. It is clear that one would want to *"minimize"* average lateness, rather than *'maximize'* it.

Figure 5.6 depicts a schematic diagram of these flow patterns.

The n-job one-machine problem we had earlier is classified as $n/1/F/\overline{F}$.

The number of different combinations in which the jobs and machines can be arranged rises exponentially with n and m, and depend on the job shop type as given in Table 5.1.

Table 5.1. **Number of combinations of possible sequences**

Job Shop Type	Number of Schedules	Example: $n = 5$, $m = 3$
G, no conditions	$> (n!)^m \cdot m!$	>10,368,000
F, all machine orders	$(n!)^m \cdot m!$	10,368,000
F, one machine order	$(n!)^m$	1,728,000
P, all machine orders	$(n!) \cdot (m!)$	720
P, one machine order	$(n!)$	120

We notice that even for small problems the number of combinations can be very high. Using a computer for medium-size problems is possible, but for larger

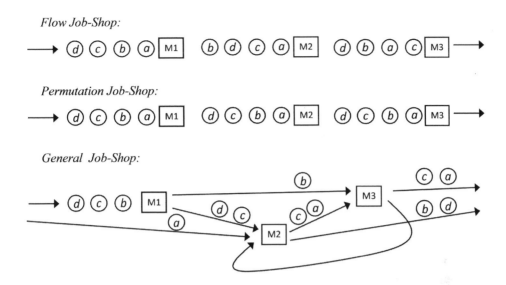

Figure 5.6. Classification of flow patterns in a job shop.

problems, even a computer will require impractical durations to generate answers. Scheduling algorithms can be effective in determining optimal solutions; plus they provide insight and understanding of the solution method. This experience can be invaluable in analyzing and handling day-to-day scheduling problems.

5.5 JOHNSON'S ALGORITHM

Johnson's algorithm is used for the n-job two-machine scheduling problem and is sometimes referred to as $n/2/F/C_{max}$ classification, but its correct classification is $n/2/P/C_{max}$.

The letter P (permutation) in $n/2/P/C_{max}$ indicates that jobs go in one direction from the first machine to the second, and the queue order (sequence) in front of both machines is the same. So one overall sequence is determined and used throughout. The restriction that jobs go from one machine to the next is often referred to as the "*technological order*", meaning that the physical properties of the job are altered in the given order, which is typical of most industrial scenarios.

Application of Johnson's algorithm is easy and straightforward, and a numerical example is the best way to demonstrate it.

Assume the information in Table 5.2 for a four-job problem. The numerical entries in the table indicate the processing times in the unit used.

Table 5.2. Data for Johnson's two-machine example

Job ID	Machine 1 (M_1)	Machine 2 (M_2)
a	4	9
b	8	3
c	7	6
d	9	5

The Algorithm

The steps of the algorithm are as follows:

- Scan the entire processing times and select the smallest value.
- If the value is in the M_1 column, place the corresponding job first in the lineup.
- If the value is in the M_2 column, place the corresponding job last.
- Cross out the job on the table.
- Repeat the above steps until all jobs have been assigned a position in the lineup.
- If a tie exists (that is, there are two equal values to choose from), break the tie arbitrarily, but if relevant choose to your advantage.

In following the steps of the algorithm, Figure 5.7 shows the sequence of forming the lineup and the order of crossing out the table entries.

The Logic behind Johnson's Algorithm

The rationale of Johnson's algorithm can be stated as:

- Line up the jobs with smaller first operation in front (head) of the queue so that the second machine is fed as soon as possible.
- Line up the jobs with smaller second operation in back (tail) of the queue so that the second machine stays busy with longer operations.

5.6 JOHNSON'S EXTENDED ALGORITHM

Johnson's n-job two-machine method can be extended to the n-job three-machine, $n/3/P/C_{max}$, problem under the following conditions:

$$either \quad \min t_{i, M_1} \geq \max t_{i, M_2}$$
$$or \quad \min t_{i, M_3} \geq \max t_{i, M_2}$$

Jobs	M1	M2	Crossing out order
a	④	9	2nd
b	8	③	1st
c	7	6	last remaining
d	9	⑤	3rd

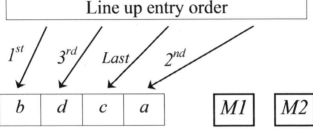

Figure 5.7. **Steps of Johnson's algorithm for the example.**

where $t_{i,j}$ stands for processing time of job i on machine j. In other words, Johnson's method can be applied if the second machine is completely dominated by either of the first or third machines. In this algorithm the three machines are converted into a two dummy or virtual machines and then Johnson's two-machine algorithm is applied.

Example 5.1

Data for processing of four jobs on three machines are given in Table 5.3. Using Johnson's extended algorithm, determine the job processing order for minimum completion time.

Table 5.3. **Data for Johnson's three-machine example**

Job ID	M_1	M_2	M_3
a	3	1	6
b	5	4	5
c	4	2	8
d	2	3	7

Solution

First, we examine to see whether any of the conditions is present:

$$\min t_{i, M_1} = 2, \min t_{i, M_3} = 5, \max t_{i, M_2} = 4$$

Since $\min t_{i, M_3} > \max t_{i, M_2}$, the algorithm can be applied. We now generate the two dummy or virtual machines and the sum the processing times as follows in Table 5.4.

Table 5.4. Dummy machines for Johnson's three-machine example

Job ID	Dummy Machine 1 $M_1 \oplus M_2$	Dummy Machine 2 $M_2 \oplus M_3$
a	3+1 = 4	1+6 = 7
b	5+4 = 9	4+5 = 9
c	4+2 = 6	2+8 = 10
d	2+ 3 = 5	3+7 = 10

The solution is usually shown as in Figure 5.8.

Figure 5.8. Solution of Johnson's extended algorithm for the example.

5.7 JACKSON'S ALGORITHM

Jackson's algorithm is based on Johnson's algorithm and is used for n-job two-machine situations where subsets of jobs have different processing orders, and it is desired to minimize the maximum completion time. The problem is defined as $n/2/G/C_{max}$.

The following explains the algorithm.

First, the pool of n jobs to be processed on two machines M_1 and M_2 is divided into four groups:

1. $\{AB\}$: Jobs that go through M_1 then M_2.
2. $\{BA\}$: Jobs that go through M_2 then M_1.

3. {*A*}: Jobs that go through M_1 only.

4. {*B*}: Jobs that go through M_2 only.

Recall that in the classification code, C_{max} means "minimize the largest or the maximum completion time." This is always the completion time of the last job in any sequence.

Depending on the case, one or two of the above groups may yield a null-set, meaning that they will have no members.

Steps of Jackson's Algorithm

1. Divide the pool of jobs into four groups as was just shown.

2. Apply Johnson's algorithm to groups {*AB*} and {*BA*}.

3. Keep the jobs in the groups {*A*} and {*B*} as they are, although they can be sequenced if there is a "practical benefit" in doing so. Timewise, however, this will not affect the algorithm or the maximum completion time.

4. Line up the four groups as shown in Figure 5.9.

The schematic diagrams of Figure 5.9 is interpreted as: once any of the sequenced groups {*AB*} and {*BA*} completes its first operation, it immediately moves to its second operation.

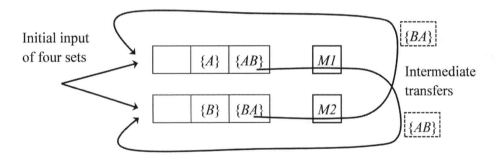

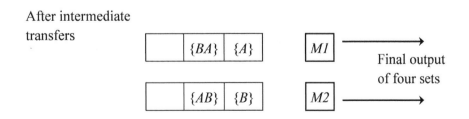

Figure 5.9. Jackson's algorithm.

The Rationale behind Jackson's Algorithm

Let's first recall the rationale supporting Johnson's algorithm:

- Line up the jobs with smaller first operation in front of the queue so that the second machine is fed as soon as possible.
- Line up the jobs with smaller second operation in back of the queue so that the second machine stays busy with longer operations.

These ideas are used twice in the Jackson's algorithm as follows:

- The sequenced group $\{AB\}$ in front of M_1 is meant to start feeding M_2 as quickly as possible.
- Similarly, the ordered group $\{BA\}$ in front of M_2 is meant to start feeding M_1 as quickly as possible.
- If the processing times of group $\{AB\}$ are such that M_1 completes it early, then there is group $\{A\}$ to maintain the machine M_1 busy until group $\{BA\}$ arrives from M_2.
- Likewise, if the processing times of group $\{BA\}$ are such that M_2 completes it early, then there is group $\{B\}$ to maintain the machine M_2 busy until group $\{AB\}$ arrives from M_1.

The overall Rationale is to achieve as many concurrent operations as possible on the two machines. The concept is smart and plausible.

With the sequence and all the processing times known, one can either mathematically, using a computer, or by drawing a Gantt chart determine the "completion" time. That is, the completion time of the last job, which, depending on the processing times, can occur on either of the machines.

5.8 AKERS' ALGORITHM

This is a practical scheduling method for two-job m-machine cases classified as $2/m/z/F_{max}$. The letter z in the classification is explained shortly. Akers' method is essentially a graphical method using a two-axes coordinate system. Figure 5.10 represents a problem and its solution. One job is assigned to each axis. Both axes must be drawn to the same scale. The alphabetic letters along each axis indicate the machine name and the stated order of processing, and the marked divisions for each machine along the axes are proportional to the processing times of the respective job.

A schedule is specified by moving (that is, graphically drawing segments of a straight line) in increasing time from zero through conflict-free areas or zones along both axes to a point akin to (x, y), where x is the total processing time of Job1 on its respective axis, ánd y is the total processing time of Job2 on its respective axis. Note that neither coordinate is the flow or completion time. A conflict area is where one machine is needed by both jobs at the same time. Such conflict areas are also called the "forbidden zones," and the line being drawn must negotiate around such zones, as demonstrated in Figure 5.10.

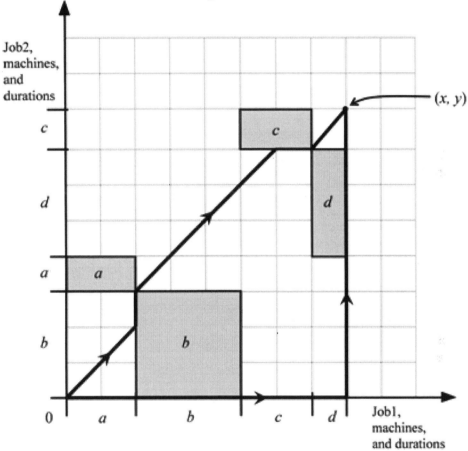

Figure 5.10. Akers' graphical algorithm.

In the classification $2/m/z/F_{max}$, the letter z can be one of the following:

    G — if the machine order is not the same of both axes

    F — if the machine order is the same for both axes, and the moving path crosses from under one forbidden zone to above another, or vice versa, at least once

P — if the machine order is the same of both axes, and the moving path is either entirely below or entirely above all forbidden zones

Regardless of the case, it has become customary to refer to the problems solved by Akers' algorithm as $2/m/F/F_{max}$. A good schedule is specified by moving in a forty-five-degree line as much as possible while circumventing the forbidden zones by moving either horizontally or vertically. It is essential that both axes have the same scale. A forty-five-degree line signifies "concurrent" operation of two machines, effectively leading to reduced flow or completion time. The optimum schedule may not be readily reached even with a good amount of forty-five-degree movement, and often minor tinkering and tweaking in the moving path is necessary to find the optimum schedule.

In Figure 5.10 the "mostly forty-five-degree heavy line" is the optimum solution as far as the order of the processes is concerned. The solution may be specified as $2/4/G/F_{max}$ since the order of machines is not the same for both jobs. The heavy horizontal line plus the heavy vertical line is a possible schedule, but it is not an optimum solution.

The completion or flow time for the set of jobs is given by either of the following equations:

$$C_{max} = F_{max} = Total\ processing\ time\ of\ \text{Job 1}$$

$$+\ \text{the time while Job 1 is waiting and Job 2 is processed}$$

$$= \sum_{j=1}^{m} t_{1,j} + \sum_{\substack{of\ schedule\ line}} (Vertical\ segments)$$

or

$$C_{max} = F_{max} = Total\ processing\ time\ of\ \text{Job 2}$$

$$+\ \text{the time while Job 2 is waiting and Job 1 is processed}$$

$$= \sum_{j=1}^{m} t_{2,j} + \sum_{\substack{of\ schedule\ line}} (Horizontal\ segments)$$

Since processing times are known and fixed, C_{max} will be minimum when the sum of horizontal lines and the sum of vertical lines are the minimum. These will occur concurrently, and the results will be the same.

Example 5.2

Two jobs are to be processed on four machines *a*, *b*, *c*, and *d*. The processing order and processing times are given in Table 5.5. Using Akers' algorithm:

i)   Find the best processing schedule and calculate the maximum flow time.

ii) Draw the respective Gantt chart.

Table 5.5. **Data of example for Akers' algorithm**

Jobs	Processing Order (Left to Right), Processing Time			
1	*d*, 3	*c*, 2	*b*, 2	*a*, 4
2	*a*, 2	*c*, 3	*b*, 4	*d*, 1

Solution

The graphical solution is given in Figure 5.11. Two schedule paths are shown. Path-1 (solid line) may appear as the best solution as we proceed in a forty-five-degree line, and once we encounter a forbidden zone, circumvent around it and continue.

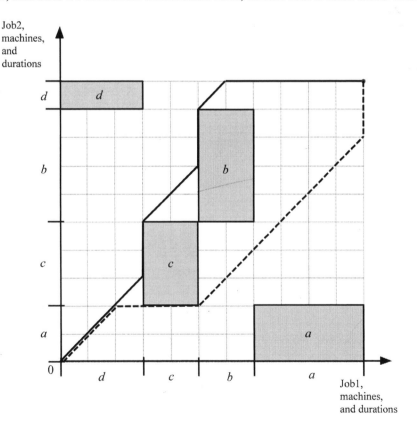

Figure 5.11. **Akers' solution for the example.**

However, as stated earlier, some tweaking may lead to a better solution. Path-2 (dashed line) is one such solution, where before running into a forbidden zone a decision is made to temporarily delay the start of processing of Job2 on Machine c.

The maximum flow time for path-1 is:

$$F_{max} = \sum_{j=1}^{m} t_{1,j} + \sum_{\substack{\text{of schedule line}}} (\textit{Vertical segments}) = 11 + 4 = 15$$

Whereas, for path 2:

$$F_{max} = \sum_{j=1}^{m} t_{1,j} + \sum_{\substack{\text{of schedule line}}} (\textit{Vertical segments}) = 11 + 2 = 13$$

The Gantt chart for Path 2 is shown in Figure 5.12.

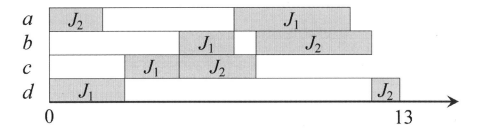

Figure 5.12. Gantt chart for the Akers' solution.

5.9 THE BRANCH AND BOUND METHOD

The branch and bound algorithm is a more general method for job shop scheduling problems. When the conditions of the Johnson's extended algorithm for n-job three-machine case do not hold the branch and bound algorithm can be used. The algorithm is developed and applied through a treelike structure where each branching node is a partial (incomplete) sequence of the jobs. Essentially, it can be formulated for any number of machines, but it will be prohibitive and involved when the number of machines exceeds five because the number of nodes and branches become unmanageable. Here we begin by demonstrating the concept behind the branch and bound algorithm with the use of a generic four-job, two-machine case; then present the formulation and example for a three-machine case; and, finally, show the formulation for the four-machine case. Most industrial and manufacturing situations rarely have five or more operations in succession requiring scheduling.

5.9.1 Lower Bounds for Completion Time

A lower bound means the lowest possible (not exact) completion time that we can determine from partially processed information. To illustrate this, let's begin with a simple four-job, two-machine problem. The Gantt chart for an arbitrarily chosen sequence will have two possible cases, and these cases are shown in Figure 5.13.

Now we can write two lower bounds on the possible shortest time by which all four jobs can be completed:

$$B_1 = \sum_{i=1}^{n} t_{i, M_1} + t_{n, M_2}$$

The equation for B_1 indicates the reality that the last job cannot start its process on M_2 until it is completed on M_1.

$$B_2 = t_{1, M_1} + \sum_{i=1}^{n} t_{i, M_2}$$

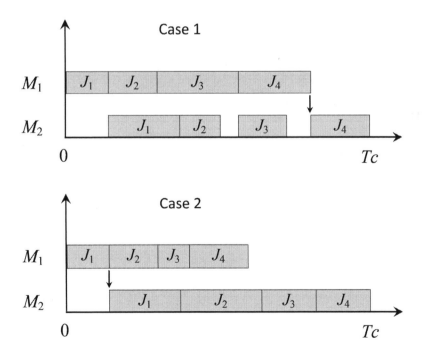

Figure 5.13. **Gantt chart cases for the branch and bound concept.**

Likewise, equation B_2 indicates the fact that none of the jobs can be processed on M_2 until Job1 is completed on M_1.

We can also look at the situation in a different way. Contrasting the above two equations against Case 1 Gantt chart of Figure 5.13, we see that B_1 gives the correct completion time and covers the duration continuously from time zero to time Tc. Equation B_2 misses the gaps (the idle times on M_2) and consequently does not give a correct completion time. We note that between the two equations, B_1 has a larger value. If we now match the above equations with Case 2 Gantt chart of Figure 5.13, we notice that now B_2 gives the correct completion time, and B_1 misses the waiting time of Job4 between M_1 and M_2. In this case, B_2 has a larger value.

No other Gantt chart case can be drawn. So, whatever the case, we can say that the correct bound on the completion time is given by the following equation:

$$Lower\ Bound = LB = \max(B_1, B_2)$$

This means that if we determine the lower LB for a particular instance, the actual completion time, Tc, will be greater or equal to LB.

The lower bounds can be readily extended to a three-machine case, and the equations will be:

$$B_1 = \sum_{i=1}^{n} t_{i, M_1} + t_{n, M_2} + t_{n, M_3}$$

$$B_2 = t_{1, M_1} + \sum_{i=1}^{n} t_{i, M_2} + t_{n, M_3}$$

$$B_3 = t_{1, M_1} + t_{2, M_2} + \sum_{i=1}^{n} t_{i, M_3}$$

With similar reasoning as we had for two-machine problem, we can say that the correct bound on the completion time for the three-machine problem is given by the following equation:

$$Lower\ Bound = LB = \max(B_1, B_2, B_3)$$

5.9.2 Branch and Bound Algorithm

In order to make the equations amenable for branching and mathematically suitable for a computer code, we redefine the above equations in a manner that we distinguish between position-assigned subset of jobs, J_a, and position-unassigned remaining subset of jobs, J_u, in any node of the branching tree. Then the lower bound on the completion time of all the schedules that begin at a particular node with the position-assigned sequence J_a is given by:

$$LB(J_a) = \max \left\{ \begin{array}{c} Time\,1(J_a) + \sum_{Ju} t_{i,\,M_1} + \min_{Ju} (t_{i,\,M_2} + t_{i,\,M_3}) \\ Time\,2(J_a) + \sum_{Ju} t_{i,\,M_2} + \min_{Ju} (t_{i,\,M_3}) \\ Time\,3(J_a) + \sum_{Ju} t_{i,\,M_3} \end{array} \right.$$

where

➢ *Time 1*, *Time 2*, and *Time 3* are times at which machines M_1, M_2, and M_3 complete processing the last job in the assigned sequence J_a.

➢ t_{ij} is the processing time of job $i = 1, 2, ..., n$ on machine j.

We sequentially generate the nodes and branch from the smallest lower bound to form the next set of the nodes. We repeat the selecting of the smallest lower bound from "all" the lower bounds determined until a complete sequence of the jobs is determined. A numerical example will make the process more comprehensible.

Example 5.3

We consider a four-job, three-machine permutation flow shop scheduling problem with the data given in Table 5.6. The "technological order" is M_1 to M_2 to M_3, and the table entries show the processing times. Let's determine the "optimum" schedule for C_{max}. (Note: This problem is classified as $4/3/P/C_{max}$, that is, minimize the completion time, not maximize!)

Table 5.6. Data for branch and bound example

Job ID	M_1	M_2	M_3
1	32	30	10
2	20	26	34
3	16	22	8
4	28	12	30

Solution

First, we branch from node 0 (the initial set where all jobs are unassigned) and construct nodes 1, 2, 3, and 4 as shown in Figure 5.14. The branches are generated progressively, although this figure depicts the entire assessed nodes and branches.

Individual digits inside the circles indicate the jobs that have already been assigned a sequence. For example, "31" means Job3 followed by Job1.

Using the equation in Section 5.9.2, we now commence on computing the lower bounds that are shown in Figure 5.14.

To compute the lower bounds, we need the following terms:

Time 1 (1) = 32, Time 2 (1) = 62, Time 3(1) = 72, Time 1 (2) = 20, Time 2 (2) = 46
Time 3 (2) = 80, Time 1 (3) = 16, Time 2 (3) = 38, Time 3 (3) = 46, Time 1 (4) = 28
Time 2 (4) = 40, Time 3 (4) = 70

The lower bounds are:

$$LB(1) = \max \begin{cases} 32 + 64 + 30 = 126 \\ 31 + 60 + 8 = 130 \\ 72 + 72 = 144 \end{cases} = 144$$

$$LB(2) = \max \begin{cases} 20 + 76 + 30 = 126 \\ 46 + 64 + 8 = 118 \\ 80 + 48 = 128 \end{cases} = 128$$

$$LB(3) = \max \begin{cases} 16 + 80 + 40 = 136 \\ 38 + 68 + 10 = 118 \\ 46 + 74 = 120 \end{cases} = 136$$

$$LB(4) = \max \begin{cases} 28 + 68 + 30 = 126 \\ 40 + 78 + 8 = 126 \\ 70 + 52 = 122 \end{cases} = 126$$

Since $LB(4) = 126$ is the smallest lower bound we branch from node 4 to generate the lower bounds for the next level. To do this, we must compute the "*Time*" terms as follows:

$$Time\,1(41) = Time\,1(4) + t_{1,1} = 28 + 32 = 60$$

$$Time\,2(41) = \max \{Time\,1(41) + t_{12},\ Time\,2(4) + t_{12}\}$$
$$= \max \{60 + 30 = 90,\ 40 + 30 = 70\} = 90$$

$$Time\,3(41) = \max \{Time\,2(41) + t_{13},\ Time\,3(4) + t_{13}\}$$
$$= \max \{90 + 10 = 100,\ 70 + 10 = 80\} = 100$$

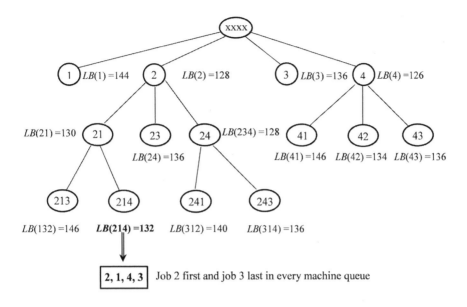

Figure 5.14. **Branch and bound complete tree.**

$$Time\,1(42) = Time\,1(4) + t_{2,1} = 28 + 20 = 48$$

$$Time\,2(42) = \max\{Time\,1(41) + t_{22},\, Time\,2(4) + t_{22}\}$$
$$= \max\{48 + 26 += 74,\, 40 + 26 = 66\} = 74$$

$$Time\,3(42) = \max\{Time\,2(42) + t_{23},\, Time\,3(4) + t_{23}\}$$
$$= \max\{74 + 34 = 108,\, 70 + 34 = 104\} = 108$$

$$Time\,1(43) = Time\,1(4) + t_{31} = 28 + 16 = 44$$

$$Time\,2(43) = \max\{Time\,1(43) + t_{32},\, Time\,2(4) + t_{32}\}$$
$$= \max\{44 + 22 = 66,\, 40 + 22 = 62\} = 66$$

$$Time\,3(43) = \max\{Time\,2(43) + t_{33},\, Time\,3(4) + t_{33}\}$$
$$= \max\{66 + 8 = 74,\, 70 + 8 = 78\} = 78$$

The lower bounds are determined as:

$$LB(41) = \max \begin{Bmatrix} 60 + 36 + 30 = 126 \\ 90 + 48 + 8 = 146 \\ 100 + 42 = 142 \end{Bmatrix} = 146$$

$$LB(42) = \max \begin{Bmatrix} 48 + 48 + 30 = 126 \\ 74 + 52 + 8 = 134 \\ 108 + 18 = 126 \end{Bmatrix} = 134$$

$$LB(43) = \max \begin{Bmatrix} 44 + 52 + 40 = 136 \\ 66 + 56 + 10 = 132 \\ 78 + 44 = 122 \end{Bmatrix} = 136$$

Because $LB(2) = 128$ is now the smallest lower bound, we return and branch from node 2. First, we compute the following required "*Time*" quantities:

$$Time1(21) = Time1(2) + t_{11} = 20 + 32 = 52$$

$$Time2(21) = \max \{Time1(21) + t_{12}, Time2(2) + t_{12}\}$$
$$= \max \{62 + 30 = 82, 46 + 30 = 76\} = 82$$

$$Time3(21) = \max \{Time2(21) + t_{13}, Time3(2) + t_{13}\}$$
$$= \max \{82 + 10 = 92, 40 + 10 = 90\} = 92$$

$$Time1(23) = Time1(2) + t_{31} = 20 + 16 = 36$$

$$Time2(23) = \max \{Time1(23) + t_{32}, Time2(2) + t_{32}\}$$
$$= \max \{36 + 22 = 58, 46 + 22 = 68\} = 68$$

$$Time3(23) = \max \{Time2(23) + t_{33}, Time3(2) + t_{33}\}$$
$$= \max \{68 + 8 = 76, 80 + 8 = 88\} = 88$$

$$Time1(24) = Time1(2) + t_{41} = 20 + 28 = 48$$

$$Time\,2\,(24) = \max\left\{Time\,1(24) + t_{42},\; Time\,2\,(2) + t_{42}\right\}$$
$$= \max\left\{48 + 12 = 60,\; 46 + 12 = 58\right\} = 60$$

$$Time\,3\,(24) = \max\left\{Time\,2(24) + t_{43},\, Time\,3\,(2) + t_{43}\right\}$$
$$= \max\left\{60 + 30 = 90,\; 80 + 30 = 110\right\} = 110$$

The lower bounds are determined as:

$$LB(21) = \max\left\{\begin{array}{l} 52 + 44 + 30 = 126 \\ 82 + 34 + 8 = 124 \\ 92 + 38 = 130 \end{array}\right\} = 130$$

$$LB(23) = \max\left\{\begin{array}{l} 36 + 60 + 40 = 136 \\ 68 + 42 + 10 = 120 \\ 88 + 40 = 128 \end{array}\right\} = 136$$

$$LB(24) = \max\left\{\begin{array}{l} 48 + 48 + 30 = 126 \\ 60 + 52 + 8 = 120 \\ 110 + 19 = 128 \end{array}\right\} = 128$$

Since $LB(24) = 128$ is the smallest lower bound, we branch from node 24 and compute the following terms:

$$Time\,1(241) = Time\,1(24) + t_{11} = 48 + 32 = 80$$

$$Time\,2\,(241) = \max\left\{Time\,1(241) + t_{12},\, Time\,2(24) + t_{12}\right\}$$
$$= \max\left\{80 + 30 = 110,\; 60 + 30 = 90\right\} = 110$$

$$Time\,3\,(241) = \max\left\{Time\,2(241) + t_{13},\, Time\,3(24) + t_{13}\right\}$$
$$= \max\left\{110 + 10 = 120,\; 110 + 10 = 120\right\} = 120$$

$$Time\,1(243) = Time\,1(24) + t_{31} = 48 + 16 = 64$$

$$Time\,2\,(243) = \max\left\{Time\,1(243) + t_{32},\, Time\,2(24) + t_{32}\right\}$$
$$= \max\left\{64 + 22 = 86,\; 60 + 22 = 82\right\} = 86$$

$$Time\,3\,(243) = \max\left\{Time\,2(243) + t_{33},\, Time\,3(24) + t_{33}\right\}$$
$$= \max\left\{86 + 8 = 94,\; 110 = 8 = 118\right\} = 118$$

The lower bounds are:

$$LB(241) = \max \left\{ \begin{array}{l} 80 + 16 + 30 = 126 \\ 110 + 22 + 8 = 140 \\ 120 + 8 = 128 \end{array} \right\} = 140$$

$$LB(243) = \max \left\{ \begin{array}{l} 64 + 32 + 40 = 136 \\ 86 + 30 + 10 = 126 \\ 118 + 10 = 128 \end{array} \right\} = 136$$

Now the smallest lower bound is the previously calculated $LB(21) = 130$; thus, we must return to node 21 and continue from there. We also need the following terms:

$$Time1(213) = Time1(21) + t_{31} = 52 + 16 = 68$$

$$Time2(213) = \max \{Time1(213) + t_{32}, Time2(21) + t_{32}\}$$
$$= \max \{68 + 22 = 90, 82 + 22 = 104\} = 104$$

$$Time3(213) = \max \{Time2(213) + t_{33}, Time3(21) + t_{33}\}$$
$$= \max \{104 + 8 = 112, 92 + 10 = 102\} = 112$$

$$Time1(214) = Time1(21) + t_{41} = 52 + 28 = 80$$

$$Time2(214) = \max \{Time1(214) + t_{42}, Time2(21) + t_{42}\}$$
$$= \max \{680 + 12 = 92, 82 + 12 = 94\} = 94$$

$$Time3(214) = \max \{Time2(214) + t_{43}, Time3(21) + t_{43}\}$$
$$= \max \{94 + 30 = 124, 92 + 30 = 112\} = 124$$

The lower bounds become:

$$LB(213) = \max \left\{ \begin{array}{l} 68 + 28 + 42 = 138 \\ 104 + 12 + 30 = 146 \\ 112 + 30 = 142 \end{array} \right\} = 140$$

$$LB(214) = \max \left\{ \begin{array}{l} 80 + 16 + 30 = 126 \\ 94 + 22 + 8 = 124 \\ 124 + 8 = 132 \end{array} \right\} = 132$$

At this point, we see that the smallest overall lower bound on the tree is $LB(214) = 132$. Therefore, the optimal sequence is {*2, 1, 4, 3*}, where job 2 is processed first, and job 3 the last. The solution is shown graphically in the Gantt chart of Figure 5.15.

The equivalent Gantt chart using Johnson's *n*-job three-machine algorithms is shown in Figure 5.16. Since machine M_2 is not dominated by either of machines M_1 or M_3, the solution sequence {*1, 3, 4, 2*} is suboptimal.

The formula set for the calculation of the lower bound for a four-machine permutation shop is as follows:

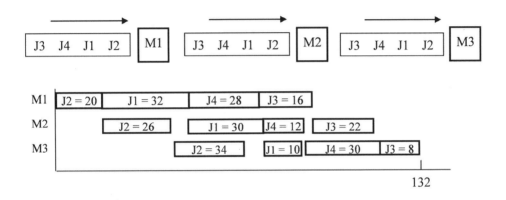

Figure 5.15. Gantt chart for the branch and bound example.

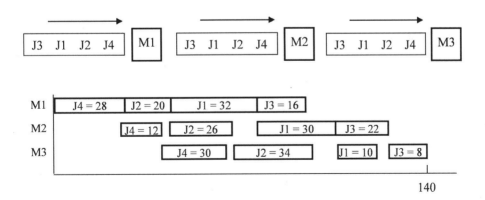

Figure 5.16. Johnson's solution for the branch and bound example.

$$LB(J_r) = \max \begin{cases} L1 = Time\,1(J_a) + \sum_{J_u}\left(t_{i,\,M_1}\right) + \min_{J_u}\left(t_{i,\,M_2} + t_{i,\,M_3} + t_{i,\,M_4}\right) \\[2mm] L2 = Time\,2(J_a) + \sum_{J_u}\left(t_{i,\,M_2}\right) + \min_{J_u}\left(t_{i,\,M_3} + t_{i,\,M_4}\right) \\[2mm] L3 = Time\,3(J_a) + \sum_{J_u}\left(t_{i,\,M_3}\right) + \min_{J_u}\left(t_{i,\,M_4}\right) \\[2mm] L4 = Time\,4(J_a) + \sum_{J_u}\left(t_{i,\,M_4}\right) \end{cases}$$

Expectedly, there will be some significant calculations involved, but they are still many orders of magnitude smaller than generating a complete set of combinations. Typically, since many of the day-to-day job shop scheduling problems call for minimizing the completion time, and the branch and bound always produces an optimal solution, it will be a worthwhile effort to code the method in a computer so that it becomes a readily reusable program.

5.10 MATHEMATICAL SOLUTIONS

In the introduction to this chapter, we indicated that virtually thousands of solutions, concepts, and algorithms have been developed for a multitude of theoretical and practical scheduling problems. However, not every solution can be found in a book or on other media, and not every problem has been addressed in the past. New situations, in terms of processing conditions or the criteria for optimization, can arise anytime. In some cases, if the problem can be cast in a simple mathematical form, then it may be possible to develop a mathematical solution. In this section, we review two such cases through parametric examples.

Example 5.4

In a post production line, units of products must be subjected to n sequential tests. The tests are independent, meaning that they can be conducted in any order. Each unit is tested at one station; if it passes, it is forwarded to the next station, and if it fails, it is rejected. There is a cost associated with each test, C_i, and from historical data, the probability of passing each test, p_i, is known. A schematic diagram of the test stations is shown in Figure 5.17.

The expected cost per unit of product for all tests for any arbitrary sequence is:

$$C = C_1 + p_1 C_2 + p_1 p_2 C_3 + \cdots + \left(p_1 p_2 \,\cdots\, p_{n-1}\right) C_n$$

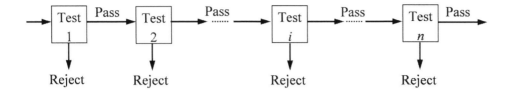

Figure 5.17. **Schematic representation of the test line.**

It may not be apparent at the first instant, but it may be possible to determine a least-cost sequence for the entire line. In this case, we can minimize the total cost by ordering the tests according to the following ranking:

$$\frac{C_{[1]}}{1-p_{[1]}} \le \frac{C_{[2]}}{1-p_{[2]}} \le \cdots \le \frac{C_{[n]}}{1-p_{[n]}}$$

Proof

We redraw the test line in an expanded form shown in Figure 5.18(a) and assume that the tests have been arranged in the ideal least-cost order.

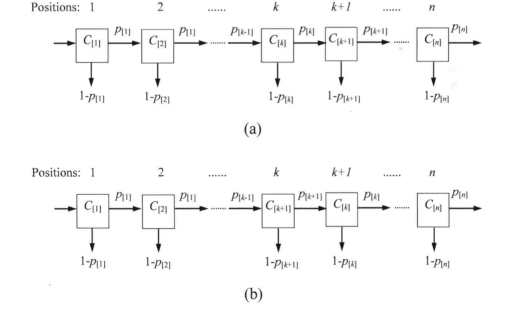

Figure 5.18. **Expanded representations of the test line.**

We now rewrite the cost function and explicitly express the cost terms for the test stations in positions $[k]$ and $[k + 1]$. We show the sum of the test costs for all the stations before position $[k]$ as A_1, and the sum of test costs for all the stations after position $[k + 1]$ as A_2. Then the total cost per unit of product will be:

$$C = A_1 + P_{[1]} \cdots \cdots P_{[k-2]} P_{[k-1]} C_{[k]} + P_{[1]} \cdots \cdots P_{[k-2]} P_{[k-1]} P_{[k]} C_{[k+1]} + A_2$$

We now switch around the stations in the positions $[k]$ and $[k + 1]$ as shown in Figure 5.18(b). This will be a non ideal order, and the total cost per unit of product will be:

$$C' = A_1 + P_{[1]} \cdots \cdots P_{[k-2]} P_{[k-1]} C_{[k+1]} + P_{[1]} \cdots \cdots P_{[k-2]} P_{[k-1]} P_{[k+1]} C_{[\ ]} + A_2$$

Subtracting one equation from the other, the terms A_1 and A_2 are eliminated, and knowing that the cost for the nonideal case could be greater or at most the same, we obtain:

$$C - C' = P_{[1]} \cdots \cdots P_{[k-2]} P_{[k-1]} C_{[k]}$$
$$+ P_{[1]} \cdots \cdots P_{[k-2]} P_{[k-1]} P_{[k]} C_{[k+1]} - P_{[1]} \cdots \cdots P_{[k-2]} P_{[k-1]} C_{[k+1]}$$
$$- P_{[1]} \cdots \cdots P_{[k-2]} P_{[k-1]} P_{[k+1]} C_{[k]} \leq 0$$

Crossing out the common terms (from $p_{[1]}$ to $p_{[k-1]}$) we have:

$$C_{[k]} + P_{[k]} C_{[k+1]} - C_{[k+1]} - P_{[k+1]} C_{[k]} \leq 0$$

After factorizing, we obtain:

$$C_{[k]} \left(1 - P_{[k+1]}\right) \leq C_{[k+1]} \left(1 - P_{[k]}\right)$$

Transposing the terms we arrive at:

$$\frac{C_{[k]}}{\left(1 - P_{[k]}\right)} \leq \frac{C_{[k+1]}}{\left(1 - P_{[k+1]}\right)}$$

This relation implies that the least-cost testing sequence will be achieved if the ranking term $\dfrac{C_{[k]}}{\left(1 - P_{[k]}\right)}$ is calculated for all the tests, and the tests are carried out in the increasing order of the value of the ranking term.

Example 5.5

We repeat Example 5.4 with an embellishment such that each unit is tested at one station and if it passes it is forwarded to the next station, but if it fails it is first repaired at a cost of R_i and then routed to the next test station. As before, there is a cost associated with each test, denoted by C_i, and from historical data the probability of passing each test, p_i, is known. A schematic representation of this test line is shown in Figure 5.19. We would like to determine what testing sequence might lead to the least-cost test line.

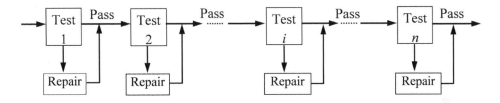

Figure 5.19. Schematic representation of the test line.

Solution

Following the steps we used in Example 5.4, we redraw a test line as shown in Figure 5.20(a), write the cost function and explicitly express the cost terms for test stations in positions $[k]$ and $[k+1]$.

We show the sum of the test and repair costs for all the stations before position $[k]$ as B_1, and the sum of test and repair costs for all the stations after position $[k+1]$ as B_2. Then the total cost per unit of product where we assume the tests are arranged in the ideal least cost sequence is:

$$C = B_1 + C_{[k]} + \left(1 - p_{[k]}\right).\ R_{[k]} + C_{[k+1]} + \left(1 - p_{[k+1]}\right).\ R_{[k+1]} + B_2$$

Now we switch around the tests in the positions $[k]$ and $[k+1]$ as shown in Figure 5.20(b). This would then be a nonideal order, and the total cost per unit of product will be:

$$C' = B_1 + C_{[k+1]} + \left(1 - p_{[k+1]}\right).R_{[k+1]} + C_{[k]} + \left(1 - p_{[k]}\right).R_{[k]} + B_2$$

Subtracting one equation from the other and noting that both equations have exactly the same terms, all are eliminated, and knowing that the cost for the nonideal case could be greater or the same, we obtain:

$$C - C' = 0$$

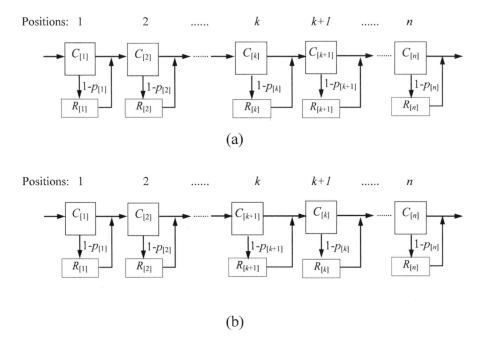

Figure 5.20. **Expanded representations of the test line.**

Therefore, in this problem, no testing sequence will minimize the test line's total cost per unit. In other words, any testing sequences will have the same total cost.

5.11 CLOSING REMARKS

Of the many tasks that an industrial engineer performs, scheduling and sequencing of operations are perhaps the most challenging. However, at the same time, scheduling is an open-ended problem and provides ample opportunity to be innovative and creative in devising solutions and methods for emerging day-to-day problems.

EXERCISES

5-1   Eight jobs are to be processed on a single machine. The processing times are given in Table 5.7. Determine the sequence that minimizes the mean flow time.

Table 5.7. Job ID and processing times for Exercise 5-1

Job ID	a	b	c	d	e	f	g	h
Processing Time	4	6	9	5	8	7	7	3

5-2 Six jobs are to be serially processed on two machines with a fixed technological order. The respective data is given in Table 5.8.

Table 5.8. Job ID and processing times for Exercise 5-2

Job ID	Machine 1 Processing Time	Machine 2 Processing Time
a	4	7
b	7	3
c	5	8
d	2	5
e	6	4
f	6	4

i) Find the job order that minimizes the completion time.

ii) Draw a scaled Gantt chart and determine the completion time.

5-3 Five jobs must go through three processes where the process order is the same for all the jobs. Find a sequence that minimizes the completion time for the set. The processing times are given in Table 5.9.

Table 5.9. Job ID and processing times for Exercise 5-3

Job ID	Process 1	Process 2	Process 3
a	10	7	8
b	7	3	6
c	8	6	7
d	9	5	5
e	8	4	8

5-4   Two jobs have to be processed on four machines. The order of processing for both jobs is $M_1, M_2, M_3,$ and M_4. The problem can be stated as $2/4/F/F_{max}$. The processing times are given in Table 5.10.

Table 5.10. Job ID and processing times for Exercise 5-4

	Machines			
Job ID	M_1	M_2	M_3	M_4
J_1	5	1	1	5
J_2	1	6	6	1

Using graph paper determine the following:

(a)   The optimum schedule and the completion (flow) time when J_1 precedes J_2

(b)   The optimum schedule and the completion (flow) time when J_2 precedes J_1

(c)   The optimum schedule and the completion (flow) time when there is no restriction

(d)   The completion time of the worst schedule, when only one job is processed at a given time

5-5   Two jobs must be processed on four machines in the order and process times shown in Table 5.11.

Table 5.11. Job ID and processing times for Exercise 5-5

Job 1		Job 2	
Machine	Time Units	Machine	Time Units
a	4	d	6
b	6	b	4
c	8	c	4
d	4	a	8

a)   Graphically show the best sequence or strategy you might use.

b)   Calculate the completion time.

c)   What will be the $n/m/A/B$ classification for the solution you have provided?

d)   Draw a properly scaled Gantt chart showing the completion time.

5-6 Table 5.12 shows the list of jobs that are all available and must be processed on a single machine. It is known that the set of jobs (c, f, i, l) is for a customer, say, #1, and the set of jobs $(a, b, d, h, j,)$ is for another customer, say, #2. The jobs in each set must be processed back-to-back (that is, together as a group and delivered to each customer). The problem can be stated as $n / m / A / B = 12 / 1 / / \overline{F}$. Where $\overline{F}$ is interpreted as "minimize the mean flow time for the two sets." In other words, determine the order of processing the sets such that the mean flow time is minimum.

Table 5.12. Job ID and processing times for Exercise 5-6

Job ID	a	b	c	d	e	f	g	h	i	j	k	l
Processing Time	2	1	5	6	3	4	1	2	5	6	1	6

5-7 Applying Jackson's algorithm, produce a sequence that would minimize the completion time for the processing of a batch of fourteen jobs on two machines with the data given in Table 5.13. The table entries are processing times where a zero time implies no operation. Draw two symbolic machines and show the job lineup in front of each machine indicating the beginning and end of the lineups.

Table 5.13. Job ID and processing times for Exercise 5-7

Job ID	First Operation on M_1	Second Operation on M_2	Job ID	First Operation on M_2	Second Operation on M_1
A	3	6	F	15	(0)
B	7	4	G	6	6
C	5	8	H	10	3
D	6	12	I	5	12
E	2	1	J	1	(0)
			K	7	8
			L	9	4
			M	14	(0)
			N	2	16

Chapter 6   FORECASTING

6.1 INTRODUCTION

Forecasting, as the name implies, is estimating a future value of a time-oriented variable. Depending on the case and intent, it can take either mathematical predicting or nonmathematical surveying forms. Our focus here is on forecasting through mathematical techniques. These forms of forecasting essentially embrace various curve-fitting concepts, and the key is on deciding which concept is the best for a given situation.

6.1.1 Pre-Forecasting Analysis

Since mathematical forecasting is based on some existing data, it could be useful if the relevance of data is ascertained and visually examined. Often we have tabular data in two or more columns. The first column is the independent variable and is usually "time," although it can be any variable, and the other column(s) will be the values of the parameter of interest observed or collected in different contexts or from different sources. A forecast will only make sense when the pairs of data exhibit some meaningful relationship. The best way to investigate this is to plot the variable of interest against the independent variable. This will not only enable one to qualitatively observe the correlation between the data but also could reveal what form of mathematical forecasting technique may be appropriate.

6.1.2 Forecasting Horizon

In Chapter 2 we discussed the topic of the planning horizon, and we divided it into three long, medium, and short ranges. Forecasting is one of the key tasks in the management planning process. It is also divided into three ranges.

Long-range or strategic forecasting involves major decisions that must be made on matters such as investment, expansion of business, and new product and services, or research and development. Sometimes these decisions are made based on governmental policies that are usually long-term, and corporations must forecast and determine how they might be affected and adapt accordingly. Long-range forecasting for five years and beyond must be looked into carefully, and some deep analysis and planning may be required. Too many external influences make long-range forecasting an arduous task that often leads to an inaccurate outcome.

Medium-range or tactical forecasting is typically from six months to five years. A frequent assessment of a medium range forecast can to some extent make up for the inaccuracies of long-range forecasting. What should be considered in medium range planning depends on the nature of the organization. Hospitals and educational institutions, producers of consumer goods, and heavy-duty machinery manufacturers may find medium-range forecasting most suitable for their planning purposes.

Short-range or operational forecasting horizon is from hours to one year or so. There are, of course, no sharp boundaries between the three forecasting horizons. A short-range forecasting horizon for an organization may be well over one year. Much of this depends highly on how often data become available to run or update a forecast, and how fast the enterprise can react to the results of a forecast. Typical examples of applications include production and personnel planning, inventory control, and space issues.

6.2 MATHEMATICAL FORECASTING METHODS

Forecasting can be done in a number of ways. In the absence of historical or relevant data, it can be accomplished without mathematical means through meetings, guessing, and seeking an expert opinion. In some cases, such as when launching a new product, past data will not be available, and, if judged relevant, the historical data of similar products may be used. For established enterprises, best forecasting can be done using their own data. There are many methods of mathematical forecasting available, and we review some fundamental techniques.

6.2.1 Linear Regression

Linear regression is a line-fitting technique that is common in analyzing experimental data to determine the relationship between two variables. Linear regression also has applications in the forecasting of a future value of a variable of interest. If the underlying observed data tend to have a linear trend, increasing or decreasing, then the linear regression is a logical choice. In this method, a straight line that "best" fits the data is mathematically determined based on what is frequently referred to as the "least squares" method.

Given a set of n paired data (x_i, y_i), where normally x_i is the independent variable and y_i is the corresponding observation, a line in the form of $\hat{y}_i = mx_i + c$ is constructed, where $\hat{y}$ is the linear estimation of the dependent variable as a function of the independent variable. The numerical values of the slope, m, and the intercept, c, are determined from the data set using the least squares method as follows.

For a point i, the error between the actual observed value and the regressed value is:

$$e_i = y_i - \hat{y}_i$$

Figure 6.1 shows the error for two sample points.

The sum of squared errors is:

$$S = \sum_{i=1}^{n} (y_i - \hat{y}_i)^2$$

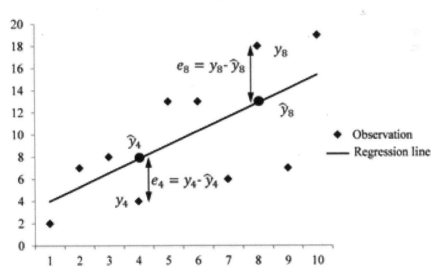

Figure 6.1. Errors between observations and the linear regression line.

If this sum is minimized, then the line passing through the data is the best fit. The parameters that can be manipulated to reposition and reorient the line are m and c. By setting the partial derivative of S with respect to m and c to zero (to minimize S), we find two simultaneous equations to determine m and c for the best fit.

Expanding the equation for S as:

$$S = \left(y_1 - \hat{y}_1\right)^2 + \left(y_2 - \hat{y}_2\right)^2 \ldots \ldots \ldots + \left(y_n - \hat{y}_n\right)^2$$

or

$$S = y_1^2 + \hat{y}_1^2 - 2y_1\hat{y}_1 + y_2^2 + \hat{y}_2^2 - 2y_2\hat{y}_2 \ldots \ldots \ldots + y_n^2 + \hat{y}_n^2 - 2y_n\hat{y}_n$$

Removing known constant terms, we obtain:

$$S' = \hat{y}_1^2 - 2y_1\hat{y}_1 + \hat{y}_2^2 - 2y_2\hat{y}_2 \ldots \ldots \ldots + \hat{y}_n^2 - 2y_n\hat{y}_n$$

Substituting from $\hat{y}_i = mx_i + c$ we have:

$$S' = \left(mx_1 + c\right)^2 - 2y_1\left(mx_1 + c\right) + \left(mx_2 + c\right)^2 - 2y_2\left(mx_2 + c\right)\ldots\ldots\ldots$$
$$+ \left(mx_n + c\right)^2 - 2y_n\left(mx_n + c\right)$$

Since we have removed the constants terms, the partial derivatives of S' will be the same as the partial derivatives of S. Expanding S', taking partial derivative with respect to m, simplifying, collecting terms into summation form, and setting the result equal to zero we obtain:

$$\frac{dS'}{dm} = m \left(\sum_{i=1}^{n} x_i^2 + c \left(\sum_{i=1}^{n} x_i\right) - \sum_{i=1}^{n} x_i y_i = 0\right.$$

Similarly:

$$\frac{dS'}{dc} = nc + m \sum_{i=1}^{n} x_i - \sum_{i=1}^{n} y_i = 0$$

Solving the above two equations simultaneously, we arrive at:

$$m = \frac{n\sum(xy) - \sum x \sum y}{n\sum x^2 - \left(\sum x\right)^2} = \frac{\sum(xy) - n\bar{x}\,\bar{y}}{\sum x^2 - n(\bar{x})^2}$$

$$c = \frac{\sum y - m\sum x}{n} = \bar{y} - m\bar{x}$$

For simplicity, the summations limits are not shown, and where:

$$\bar{x} = \frac{\sum x}{n}, \bar{y} = \frac{\sum y}{n}$$

Example 6.1

The demand for a product has been recorded for ten consecutive months as given in Table 6.1. An initial plot of the data suggests a linear trend forecasting as a suitable choice. It is desired to forecast the demand for months 11 and 12.

Table 6.1. Linear regression

Period for demand *t (months)*	Demand *D (1,000 units)*
1	5
2	7
3	6
4	7
5	9
6	8
7	9
8	12
9	11
10	13

Solution

To use the equations for the slope and the intercept, we replace x by t, and y by D. The slope and the intercept are calculated as $m = 0.82$ and $c = 4.2$. The linear regression equation is:

$$\hat{D}_t = 0.82\,t + 4.2$$

The forecasts for $t = 11$ and 12 months are:

$$\hat{D}_{11} = 13.2\ [1000\ units]$$
$$\hat{D}_{12} = 14.0\ [1000\ units]$$

A better forecast for month 12 can be obtained after the actual demand data for month 11 becomes available, and the regression line is updated. Generally, any

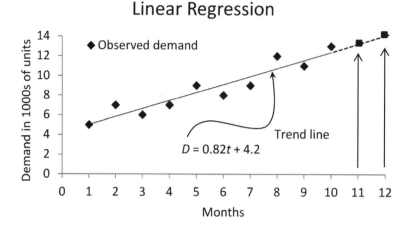

Figure 6.2. Forecasting with linear regression.

forecasting method should not be used for too many periods into the future (up to two periods is reasonable).

6.2.2 Simple Moving Average

In this method, the average of a predetermined number of most recent observations of the variable of interest is used as the forecast value of the variable for the next period. Let:

F_0 be the forecast being made for the next period
n be the number of most recent observations to be used
O_t be the observation made t periods ago
where $t = 1$ (the present period) to n (the most distant period)

Then:

$$F_0 = \frac{\sum_{t=1}^{n} O_t}{n} = \sum_{t=1}^{n} \frac{1}{n} O_t = \sum_{t=1}^{n} w\, O_t$$

For $n = 4$

$$F_0 = 0.25\, O_1 + 0.25\, O_2 + 0.25\, O_2 + 0.25\, O_4$$

We note that each observation is given in a sense an equal weight of $w = 0.25$, where the sum of the weights is unity. The key here is to determine a reasonable number of past observations to use. A few trials with error analysis and experience will help arrive at this number.

Example 6.2

The demands for last three weeks of a piece of equipment are shown in a bold, italic font in Table 6.2. A three-point moving average method is to be used to forecast future demands until a twenty-week demand data have been compiled. Determine the forecasts and forecasts error, and present the results in a plot.

Table 6.2. **Data for simple moving average**

Week	Demand	Forecast	Forecast Error
1	*12*		
2	*14*		
3	*15*		
4	12	13.67	1.67
5	13	13.67	0.67
6	12	13.33	1.33
7	16	12.33	-3.67
8	15	13.67	-1.33
9	18	14.33	-3.67
10	18	16.33	-1.67
11	17	17.00	0.00
12	19	17.67	-1.33
13	17	18.00	1.00
14	15	17.67	2.67
15	16	17.00	1.00
16	14	16.00	2.00
17	14	15.00	1.00
18	15	14.67	-0.33
19	13	14.33	1.33
20	12	14.00	2.00
21		13.33	

Figure 6.3 shows the results for this example.

Solution

We use the demands for weeks 1, 2, and 3, and determine the forecast for week 4 as given in the table. Once the actual demand for week 4 becomes available, we use

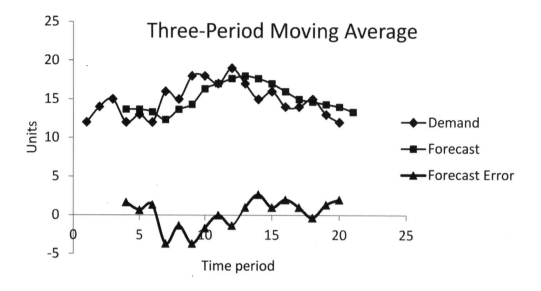

Figure 6.3. Forecasting with the simple moving average.

the observations from weeks 2, 3, and 4 to forecast the demand for week 5, and so on. The weight of each demand is: $w = \dfrac{1}{3} = 33.33\%$

6.2.3 Weighted Moving Average

The concept of "weights" gives rise to the weighted moving average method. Unlike the simple moving average method, where the observed data points have equal weights, in this approach, a predetermined number of most recent observations are each given a specific weight, w_t, where they are unequal, but the weights must sum to unity. The period t is defined as in Section 6.2.2, and the forecast is given by:

$$F_0 = \sum_{t=1}^{n} w_t O_t$$

$$\sum_{t=1}^{n} w_t = 1 = 100\%$$

The number of observed points and the value of each weight can be determined from the consideration of the problem at hand, and the use of trial and error analysis. Typically, the more recent the observation, the larger the weight associated with it.

Example 6.3

Suppose for data given in the previous example, Table 6.2, it is now desired to use the three-point ($n = 3$) weighted moving average method for the forecasts. The weight for the successive demands are $w_1 = 70\%$, $w_2 = 20\%$, and $w_3 = 10\%$, where the first is associated with the most recent demand and the last with the most distant. The initial data is given in the first three rows of Table 6.3.

Table 6.3. Weighted moving average

Week	Demand	Forecast	Forecast Error
1	12		
2	14		
3	15		
4	12	14.5	2.5
5	13	12.8	-0.2
6	12	13.0	1
7	16	12.2	-3.8
8	15	14.9	-0.1
9	18	14.9	-3.1
10	18	17.2	-0.8
11	17	17.7	0.7
12	19	17.3	-1.7
13	17	18.5	1.5
14	15	17.4	2.4
15	16	15.8	-0.2
16	14	15.9	1.9
17	14	14.5	0.5
18	15	14.2	-0.8
19	13	14.7	1.7
20	12	13.5	1.5
21		12.5	

Solution

The results from using the weighted moving average equation are also shown in Table 6.3. The observed demand and the forecasting curve are shown in Figure 6.4.

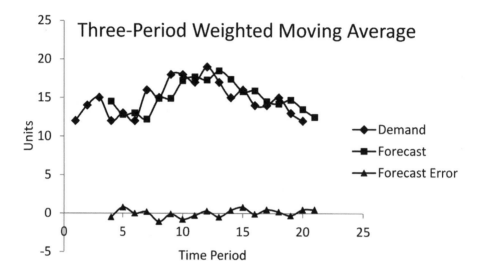

Figure 6.4. Forecasting with weighted moving average.

6.2.4 Exponential Smoothing

Exponential smoothing is a simple and yet an elaborate form of the weighted moving average. In most instances it is reasonable to assume that, as we go back in time, the importance of data for a future forecast diminishes. Exponential smoothing accomplishes this assumption readily. The main idea is to cast the problem in a recursive formulation where the forecast for any period is written in terms of the data of the preceding period plus a correction term. This can be done in two ways:

$$New\ Forecast = Old\ Forecast + \alpha\,(Latest\ obsrvation - Old\ Forecast)$$

or

$$New\ Forecast = Latest\ obsrvation + \beta\,(Old\ Forecast - Latest\ Observation)$$

Although the second equation appears to be based on factual data (the latest observation), both expressions lead to an identical result with a change of variables by defining $\beta = 1 - a$. We leave this proof as an exercise for the reader and continue our deliberations with the first option, which is more common. Here α is referred to as the smoothing constant and is typically around 0.1–0.3, but can cover the range 0.05–0.4. The old forecast can be obtained from any forecasting method. It can also be any reasonable assumption, though it might take a few forecasting periods before the forecast equation settles. The key issue here is to decide on a suitable value for the smoothing constant, and this might take a few trial and error analyses.

We now show how the exponential smoothing method works. Referring to the variables defined in Section 6.2.2, we can write:

$$F_0 = F_1 + a\left(O_1 - F_1\right)$$

or

$$F_0 = aO_1 + \left(1 - a\right) F_1$$

But F_1 can be determined in the same fashion as:

$$F_1 = aO_2 + \left(1 - a\right) F_2$$

Substituting this into the equation for F_0 we obtain:

$$F_0 = aO_1 + \left(1 - a\right)\left[aO_2 + \left(1 - a\right) F_2\right]$$

or

$$F_0 = aO_1 + a\left(1 - a\right) O_2 + \left(1 - a\right)^2 F_2$$

We can continue the recursive expression as:

$$F_2 = aO_3 + \left(1 - a\right) F_3$$

and substituting this again into the equation for F_0 we will have:

$$F_0 = aO_1 + a\left(1 - a\right) O_2 + \left(1 - a\right)^2 \left[aO_3 + \left(1 - a\right) F_3\right]$$

or

$$F_0 = aO_1 + a\left(1 - a\right) O_2 + a\left(1 - a\right)^2 O_3 + \left(1 - a\right)^3 F_3$$

The recursive substitution can be applied indefinitely leading to the series:

$$F_0 = a[O_1 + (1-a) O_2 + (1-a)^2 O_3 + \ldots + (1-a)^k O_{k+1} \ldots] + (1-a)^\infty F_\infty$$

Since $(1-\alpha) < 1$, the last symbolic term diminishes to zero. We see that every past observation has a multiplier or "weight," and as we go back in time, they are exponentially reduced by a factor of $(1-\alpha)$. It is because of this successive reduction in weights that the method is called "exponential smoothing." The weights are: $a, a(1-a), a(1-a)^2, a(1-a)^3, \ldots \ldots a(1-a)^k \ldots \ldots$

It can be shown that:

$$\sum_{i=1}^{\infty} a(1-a)^{i-1} = 1$$

which satisfies the condition that the sum of the weights must be unity. The initial compact equation that we started with:

$$F_0 = F_1 + a(O_1 - F_1)$$

and the last expanded equation:

$$F_0 = aO_1 + a(1-a) O_2 + a(1-a)^2 O_3 + \ldots + a(1-a)^k O_{k+1} \ldots$$

are essentially the same. Therefore, it makes sense to use the first equation, which requires only three pieces of information: the forecast made for the present period (the old forecast), the observation for the present period, and the smoothing constant.

The expanded exponential smoothing forecast equation seemingly and indeed mathematically has an infinite number of terms. However, the weights are concentrated in the first few terms. The weights of the subsequent terms diminish rapidly, and desirably so. Table 6.4 shows the values of the individual and their cumulative values for the first ten terms of the equation for three typical values for the smoothing constant. For example, with $\alpha = 0.2$, by term 10, 89 percent of the weight has been allocated. The remaining 11 percent covers from term 11 for the rest of the infinite series.

Table 6.4. Distribution of weights in exponential smoothing

	Individual Weight $a\left(1-a\right)^{(i-1)}$			Cumulative Weight $\sum_{j=1}^{i} a(1-a)^{(j-1)}$		
Term in the Expanded Equation, i	$\alpha = 0.1$	$\alpha = 0.2$	$\alpha = 0.3$	$\alpha = 0.1$	$\alpha = 0.2$	$\alpha = 0.3$
1	0.10	0.20	0.30	0.10	0.20	0.30
2	0.09	0.16	0.21	0.19	0.36	0.51
3	0.08	0.13	0.15	0.27	0.49	0.66
4	0.07	0.10	0.10	0.34	0.59	0.76
5	0.07	0.08	0.07	0.41	0.67	0.83
6	0.06	0.07	0.05	0.47	0.74	0.88
7	0.05	0.05	0.04	0.52	0.79	0.92
8	0.05	0.04	0.02	0.57	0.83	0.94
9	0.04	0.03	0.02	0.61	0.87	0.96
10	0.04	0.03	0.01	0.65	0.89	0.97

From Table 6.4, the data for $\alpha = 0.2$ have been reproduced as a plot, as shown in Figure 6.5.

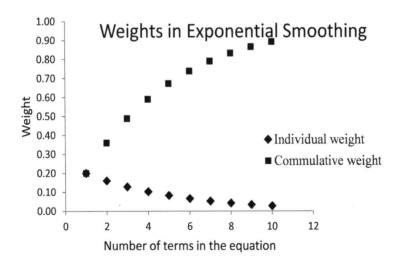

Figure 6.5. Distribution of weights in exponential smoothing.

Example 6.4

The observation for product demand and its forecast for the present week are known and shown in the bold, italic font in the first data row of Table 6.5. As the subsequent demand data become available, using exponential smoothing with a smoothing constant of $\alpha = 0.2$ forecast the demand and determine its error for the subsequent twenty weeks.

Solution

The observed demand, forecast, and forecast error are shown in Table 6.5 and plotted in Figure 6.6.

Table 6.5. Exponential Smoothing

Week	Demand	Forecast	Forecast Error
1	*15*	*14*	
2	15	14.2	-0.8
3	12	14.4	2.4
4	13	13.9	0.9
5	12	13.7	1.7
6	16	13.4	-2.6
7	15	13.9	-1.1
8	16	14.1	-1.9
9	17	14.5	-2.5
10	17	15.0	-2.0
11	16	15.4	-0.6
12	17	15.5	-1.5
13	15	15.8	0.8
14	16	15.7	-0.3
15	14	15.7	1.7
16	14	15.4	1.4
17	15	15.1	0.1
18	13	15.1	2.1
19	12	14.7	2.7
20	14	14.1	0.1
21		14.1	

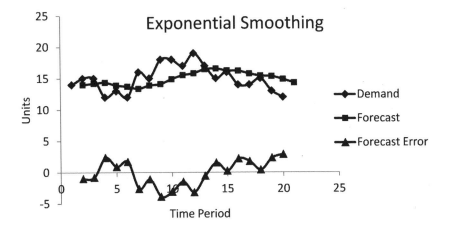

Figure 6.6. Forecasting with exponential smoothing.

6.3 MEASURE OF QUALITY OF FORECASTS

With the regression method of forecasting, a visual inspection of the plot of observed data and the fitted line will provide some indication of its suitability of its quality of fit. However, in some instances, it may be desirable to use a quantitative measure of the regression line. The coefficient of correlation is a measure that provides a useful indication of the quality of the line fit. For other forecasting methods, alternative measures are available. The coefficient of correlation and these measures are explained in the next two sections.

6.3.1 Coefficient of Correlation

The coefficient of correlation is commonly used to determine the degree of relationship between two variables. Is the case of a linear regression, it provides a measure of the closeness of the observed demand and the corresponding estimated demand. Coefficient of correlation is defined as:

$$r = \sqrt{1 - \frac{Regression\ sum\ of\ squares}{Total\ sum\ of\ squares}} = \sqrt{1 - \frac{\sum_{i=1}^{n}(y_i - \hat{y}_i)^2}{\sum_{i=1}^{n}(y_i - \bar{y})^2}}$$

where
y_i is the ith observed data point
$\bar{y}_t$ is the average of observed data points

$\hat{y}_i$ is the regressed value for i^{th} data point

$i = 1 \ldots n$

A graphical representation of the above data and errors are similar to as shown Figure 6.1.

The generally accepted rule-of-thumb interpretation of the numerical value of the coefficient of correlation is given in Table 6.6.

Table 6.6. Interpretation of the coefficient of correlation

r	Interpretation
0.9–1.0	Very High correlation
0.7–0.9	High correlation
0.4–0.7	Moderate correlation
0.2–0.4	Low correlation
0.0–0.2	Slight or no correlation

6.3.2 Analysis of Forecast Errors

Sometimes it is not clear what method of forecasting is best suited for a given situation. In such cases, it may be necessary to examine a few models and then assess the predictive accuracy or the error of the forecasts thus obtained. A number of methods are available for this purpose.

The forecast error for a particular period t is the difference between the actual observed data value, y_t, and the forecast value, $\overline{y}_t$:

$$e = y_t - \hat{y}_t$$

The sum of the errors over a number of periods is not an appropriate measure because positive and negative errors cancel out. For example, in the regression method of forecasting, the sum of the squared errors is minimum, and the sum of the errors is in fact always zero. The best methods of error analysis are based on either sum of the squared errors or the sum of the absolute value of the error. The following three methods are common in practice.

- Mean Squared Error (MSE):

$$MSE = \frac{\sum_{t=1}^{n} (y_t - \hat{y}_t)^2}{n}$$

- Mean Absolute Deviation (MAD):

$$MAD = \frac{\sum_{t=1}^{n} |y_t - \hat{y}_t|}{n}$$

- Mean Absolute Percentage Error (MAP):

$$MAP = \frac{100}{n} \sum_{t=1}^{n} \left| \frac{y_t - \hat{y}_t}{y_t} \right|$$

The first two measures have "units" of the variable of interest, and can be useful if one forecast model is being examined with a different parameter. A single measure may not provide a qualitative indication of the goodness of the forecast. The third measure, MAP, is unit-less and provides a percentage error that is a more familiar and tangible measure.

6.4 CLOSING REMARKS

Forecasting is essentially a curve-fitting process. The key is to find the form and parameters of the best fitting curve, and then through mathematical means assess the predictive accuracy of such a curve.

EXERCISES

6-1   If you are working in an organization where forecasting is a frequent activity, what is short-range forecasting is used for? Are relevant observations properly recorded? Is error analysis performed and appropriate actions are taken?

6-2   If the answer to Exercise 6-1 is negative, do you think implementing a forecasting program will be beneficial? Are you in a position to suggest such and plan and take on the task?

6-3   In follow up to Exercise 6-2, if you are able to make a suggestion, what will be your preferred choice of the forecasting method? And why?

6-4 To make better use of time in faster achieving some results, what do you think of spending some efforts to implement two forecasting methods concurrently and determine which one performs better based on error analyses?

6-5 Find the linear regression line for the data given in Table 6.7.

Table 6.7. Data for Exercise 6-5

x	y
1.6	2.6
2.0	2.9
2.4	2.0
2.7	3.1
3.5	3.4
3.8	4.1
5.0	4.5
5.4	5.5

Perform all three methods of forecast error analysis and comment on results.

6-6 Apply exponential smoothing to the data shown in Table 6.8 with a smoothing constant of $\alpha = 0.2$. Show the results in a table with column headings Observation, Forecast, and Forecast Error.

Table 6.8. Data for Exercise 6-6

20	20	20	20	24	24	24	24	24	24

6-7 Similar to Exercise 6-6, apply exponential smoothing to the data given in Table 6.9 with a smoothing constant of $\alpha = 0.2$. Show the results in a table with column headings Observation, Forecast, and Forecast Error.

Table 6.9. Data for Exercise 6-7

24	25	24	25	20	20	20	20	20	20

6-8 Carefully examine the numerical values of the errors in each of the Exercises 6-6 and 6-7. What overall observation do you make?

Chapter 7 STATISTICAL QUALITY CONTROL

7.1 INTRODUCTION

The importance of quality cannot be underestimated. The least impact resulting from a low-quality product is probably only economical, and this by itself can severely impair the financial health of any enterprise. In the worst case, inferior quality can lead to irreversible consequences, such as loss of life—which can inflict severe financial damages and endanger the existence of the enterprise. A credible enterprise must aim to produce goods and services of the right quantity and value, and also of the right quality. It is important to know the difference between *inspection* and *quality control*. Inspection is used to verify the quality of what has already been produced, whereas quality control is a process that is used to foresee, direct, and control the quality of the products yet to be produced.

When an item or product fails to meet expectations or perform its intended function, then it is associated with inferior or unacceptable quality. Technically, by "quality" we mean to what degree a measured value of a feature deviates from its stated specifications. The smaller the deviation, the better the quality.

To gauge quality, the general public and consumers rely on a company's reputation for quality and reliability, and the brand name, as they lack any systematic tools to measure quality.

The volume of the business, natural desire for profits, and sometimes the regulations in place justify, encourage and force organizations to resort to more scientific and consistent methods of assessment to produce goods of acceptable quality and constantly look for ways of improvement. Statistical quality control (SQC) is an approach that can cost-effectively help any organization to achieve these. We will show that it is not necessary for SQC to screen each and every one of the product items or components. Proven sampling techniques allow only a small part of the output to be examined, yet provide significant insight into the quality and quality trends of the entire production.*

It is not even necessary to maintain quality control efforts at all times and for all products. With good measures and a proper combination of factors of production, a process may in time indicate that active quality control is no longer needed. Nonetheless, maintaining some degree of inspection and control provides assurance that the achieved quality levels do not deteriorate.

We define factors of production as all elements that play a role in manufacturing a product or when providing a service. In manufacturing, for example, these are primarily equipment, tools, materials, operator skills, and instructions. Other factors may be such things as maintenance and ambient temperature. If proper resources and factors of production are not provided, the expectation for good quality will be unrealistic.

Ultimately, quality control efforts are inherently linked to much broader aspects of any production system and running an enterprise. These include design, performance and market indicators, cost analysis and finance, management philosophy and safety.

7.2 DETERMINING QUALITY

For small and relatively inexpensive goods, such as consumables and food, for example, personal experience is the primary means of associating a quality rating. For major products and products of high value, customers and consumers actively search various media and seek advice from others. The third, and unfortunate, means of finding out about quality is when a manufacturer or supplier issues a recall of their product as a result of defects, deficiencies, and poor quality and performance. The following is a list of a few common recall scenarios:

---

*   In this chapter we frequently use the terms *items, components, products, enterprise, firm,* and *organization* in the context of manufacturing industries to give examples and show the implementation of the quality control concepts. This, however, does not preclude in any way the application of the techniques developed to other settings such as the service industry, the health-care sector, educational institutions, and governmental organizations.

- Automobiles, practically from all manufacturers
- A range of household items, such as strollers, toys, and electronic devices
- Processed and packaged food

A simple search on the Internet with the keyword *recall* and any product or automobile brand name will invariably return a list of recalls. This is particularly prevalent in the automobile industry, as every year the manufacturers implement new designs and functionality or introduce new models that accompany new issues in quality.

When the deficiencies are significant and affect a broader base of consumers, the companies involved issue public and media announcements. Otherwise, many recalls are handled through the mail and go unnoticed by the general public. A recall can tarnish the reputation and damage the financial health of the manufacturer, and critical defects and deficiencies can lead to catastrophic consequences.

7.2.1 Why Is Quality Inconsistent?

When components or products are produced, a set of design, manufacturing, and performance specifications are necessary. It is almost impossible to state and achieve the exact specifications of the product. Even if it were possible, the manufacturing and quality control costs would be prohibitive. In fact, from the cost consideration point of view, it is not even necessary to achieve exact specifications. For example, when a shaft is being machined to support a ball bearing, we commonly state the specification as a desired target diameter plus/minus some acceptable tolerance. If a thousand such shafts are produced, they will hardly be the same.

But this is fine because we want shafts to be practically usable and economically justifiable. Inconsistencies and deviations come from numerous sources. They are occasionally random in nature, and most often the result of variations in the factors of production. Taking a machining operation, we can name the following as typical examples:

- Tool wear
- Slack in machine parts and joints
- Error in proper setting of measures and adjustments
- Variation in temperature
- Inconsistency in material properties
- Other reasons depending on the case

With automation, compensatory feedback, and improvements in the manufacturing precision of equipment and machine tools, such variations and errors have been significantly reduced. However, with higher expectations for quality, in the majority of common manufacturing processes, they will still be present.

7.3 ECONOMICS OF QUALITY

After safety has been sufficiently and legally addressed for a product or service, it is the "economics" that truly controls the level of quality. Attempts are constantly made to reduce or eliminate the effects of external factors and improve quality. Clearly, refining and improving quality exceedingly costs the manufacturer, but it also increases the value of the product. The difference between the product value and the cost of quality is the gain due to quality control efforts. Figure 7.1 shows typical trends for the cost and value of a product as a function of quality refinement. The maximum gain occurs at an optimum quality level. This is the lowest level of quality that must be attained to sustain the market. Certain manufacturers may operate above the optimum level of quality to gain reputation. This results in some brand names benefiting from a positive public opinion about their products. We should always have in mind that customers and consumers seek an acceptable level of quality that satisfies their needs in terms of cost-effectiveness. They do not need excessive quality, at least in certain aspects of the product they buy, nor are they willing to pay extra for it.

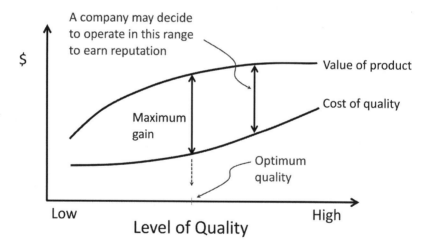

Figure 7.1. Economics of quality control.

7.4 THE ROLE OF DISTRIBUTIONS

Suppose that at a machining station holes are bored in a frame to hold an optical lens. The specification for the hole diameter is given as a target value and a tolerance. Any hole size outside the specified range will either be slack in firmly holding the lens or will be too tight and cause shattering of the lens over time.

After a large number of holes have been bored, there will certainly be variation in their respective diameter. If the variations are plotted as a frequency diagram, a possible representation might be as shown in Figure 7.2. In fact, in all similar instances, a population distribution is generated. These distributions will differ from one another even for very similar processes. Distributions are essential in devising means of implementing quality control and assessing the level of acceptable output.

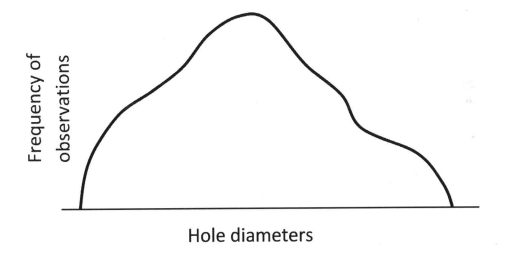

Figure 7.2. **Distribution of diameters.**

7.4.1 The Scatter of Data

Distributions have certain characteristic parameters, such as the well-known and well-used "arithmetic mean" and "standard deviation." These measures allow us to employ useful features of the distribution.

Arithmetic Mean

The average, arithmetic mean, or simply the "mean" of a distribution is given as:

$$\mu = \frac{\sum_{i=1}^{N} X_i}{N}$$

where X_i designates members of the population, and N is the size of the population.

Measures of Dispersion

The arithmetic mean alone does not provide a sufficient description of the distribution. Two distributions can have the same mean but significantly different populations. To better characterize a distribution, we need another measure that explains the spread or dispersion of the data. For this variability, two measures are available: the *range* and the *standard deviation*. The range is defined as:

$$R = X_{max} - X_{min} \geq 0$$

where X_{max} is the largest observation and X_{min} is the smallest observation within the population. The standard deviation (STD) is defined as:

$$\sigma = \sqrt{\frac{\sum_{i=1}^{N} (X_i - \mu)^2}{N}} \geq 0$$

If all members of a population have the same value, then the range and the standard deviation will be zero. As the scatter grows, both the range and standard deviation increase. As opposed to the range, which is based on two members of the population, standard deviation accounts for every member of the population. Therefore, the standard deviation is a significantly better measure. We show later that for statistical quality control purposes, and the manner in which the procedures are applied, there is a relationship between R and σ, and since R is more convenient to use, and indeed sufficient, the use of R is common and dominant.

When we have a set of reasonably related data, which is from one generating source, we can express it as a credible distribution in a mathematical form in terms of its statistical measures. One of the most common, applicable, and useful distributions is the normal distribution. In fact, the distribution of a variable, such as the diameter of machined shafts, which is influenced by several external factors, will most likely be normal, though this is not a universal rule. A machined shaft is affected by factors such as the settings on the machine tool, operator skills, sharpness of the cutting tool, and the ambient temperature.

Normal distribution is easy to understand and use. It plays a central role in statistical quality control.

7.4.2 Normal Distribution

Normal distribution is a continuous distribution, and it is entirely described in terms of its population average (arithmetic mean or simply mean) μ, and its standard deviation σ. The mathematical expression or so-called *probability density function* for normal distribution is:

$$f(X) = \frac{1}{\sqrt{2\pi}} e^{\frac{-(X-\mu)^2}{2\sigma^2}}$$

Using this function, we determine that:

$$A = \int_{-\infty}^{\infty} f(X)\, dX = 1 = 100\%$$

where A is the entire area under the probability distribution function for normal distribution which extends from minus infinity to plus infinity, signifying that it contains the entire population within. A plot of a typical normal distribution is shown in Figure 7.3.

Using the above two equations, we can obtain useful information about the nature of the population within the distribution. For example, in the second equation, if we use a lower integral limit of $\mu - 2\sigma$ and an upper integral limit of $\mu + 2\sigma$, we will obtain a numerical value of 0.954. This means that 95.4 percent of the area or population is within these limits. In other words, we can say 2σ to the

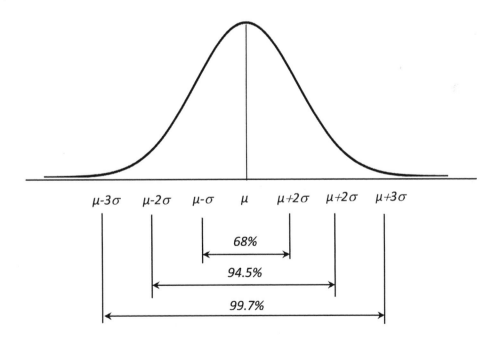

Figure 7.3. **The normal distribution.**

left and 2σ to the right of μ, or similarly a range of $\mu \pm 2\sigma$, covers 95.4 percent of the area under the normal distribution curve. In fact, using the above equations for the normal distribution, we can determine the percentage of population within any range.

Referring to our example of the bored holes, let us suppose that we have a normal distribution of the diameters with a mean of 1.502 inches and a standard deviation of 0.003 inches. Therefore, 95.4 percent of the items will have a diameter between $1.502 \pm 2(0.003)$ or 1.496 and 1.508 inches. This means that if an item is randomly selected from the population, there is 95.4 percent probability that the item diameter will be within the tolerance limits.

These calculations were carried out on the assumption that the distribution of the items (hole diameters) is normal. We also noted that normal distribution is a continuous distribution and theoretically extends from minus infinity to plus infinity. We might then question the validity of its application for quality control purposes, arguing the following:

• Our population will not contain an infinite number of items.

• The possible diameter values will in no way extend to minus infinity or plus infinity, as they will have finite positive values in the neighborhood of the stated specification.

• The diameters are measured and expressed in thousandths of an inch and therefore represent a population of discrete values.

We do acknowledge these concerns, but justify our use of the normal distribution on the following basis:

• The population of the items, in the long run, will be sufficiently large.

• 99.997 percent of the population will be within the range of $\pm 4\sigma$. This means that essentially the highly skewed tails of the distribution contain a negligible population.

• With the major portion of the population within a finite range, the density of measurements will be sufficiently high to assume continuous values.

• Even if the distribution is only close to normal, the induced errors will be negligible.

Normal Distribution Table

Instead of integrating to determine the portion of the population within a specific range, standard tables, as the one shown in Table 7.1, have been prepared that are very convenient to use.

Table 7.1. Areas under the normal distribution curve

Z	0.00	0.01	0.02	0.03	0.04	0.05	0.06	0.07	0.08	0.09
0.0	0.0000	0.0040	0.0080	0.0120	0.0160	0.0199	0.0239	0.0279	0.0319	0.0359
0.1	0.0398	0.0438	0.0478	0.0517	0.0557	0.0596	0.0636	0.0675	0.0714	0.0753
0.2	0.0793	0.0832	0.0871	0.0910	0.0948	0.0987	0.1026	0.1064	0.1103	0.1141
0.3	0.1179	0.1217	0.1255	0.1293	0.1331	0.1368	0.1406	0.1443	0.1480	0.1517
0.4	0.1554	0.1591	0.1628	0.1664	0.1700	0.1736	0.1772	0.1808	0.1844	0.1879
0.5	0.1915	0.1950	0.1985	0.2019	0.2054	0.2088	0.2123	0.2157	0.2190	0.2224
0.6	0.2257	0.2291	0.2324	0.2357	0.2389	0.2422	0.2454	0.2486	0.2517	0.2549
0.7	0.2580	0.2611	0.2642	0.2673	0.2704	0.2734	0.2764	0.2794	0.2823	0.2852
0.8	0.2881	0.2910	0.2939	0.2967	0.2995	0.3023	0.3051	0.3078	0.3106	0.3133
0.9	0.3159	0.3186	0.3212	0.3238	0.3264	0.3289	0.3315	0.3340	0.3365	0.3389
1.0	0.3413	0.3438	0.3461	0.3485	0.3508	0.3531	0.3554	0.3577	0.3599	0.3621
1.1	0.3643	0.3665	0.3686	0.3708	0.3729	0.3749	0.3770	0.3790	0.3810	0.3830
1.2	0.3849	0.3869	0.3888	0.3907	0.3925	0.3944	0.3962	0.3980	0.3997	0.4015
1.3	0.4032	0.4049	0.4066	0.4082	0.4099	0.4115	0.4131	0.4147	0.4162	0.4177
1.4	0.4192	0.4207	0.4222	0.4236	0.4251	0.4265	0.4279	0.4292	0.4306	0.4319
1.5	0.4332	0.4345	0.4357	0.4370	0.4382	0.4394	0.4406	0.4418	0.4429	0.4441
1.6	0.4452	0.4463	0.4474	0.4484	0.4495	0.4505	0.4515	0.4525	0.4535	0.4545
1.7	0.4554	0.4564	0.4573	0.4582	0.4591	0.4599	0.4608	0.4616	0.4625	0.4633
1.8	0.4641	0.4649	0.4656	0.4664	0.4671	0.4678	0.4686	0.4693	0.4699	0.4706
1.9	0.4713	0.4719	0.4726	0.4732	0.4738	0.4744	0.4750	0.4756	0.4761	0.4767
2.0	0.4772	0.4778	0.4783	0.4788	0.4793	0.4798	0.4803	0.4808	0.4812	0.4817
2.1	0.4821	0.4826	0.4830	0.4834	0.4838	0.4842	0.4846	0.4850	0.4854	0.4857
2.2	0.4861	0.4864	0.4868	0.4871	0.4875	0.4878	0.4881	0.4884	0.4887	0.4890
2.3	0.4893	0.4896	0.4898	0.4901	0.4904	0.4906	0.4909	0.4911	0.4913	0.4916
2.4	0.4918	0.4920	0.4922	0.4925	0.4927	0.4929	0.4931	0.4932	0.4934	0.4936
2.5	0.4938	0.4940	0.4941	0.4943	0.4945	0.4946	0.4948	0.4949	0.4951	0.4952
2.6	0.4953	0.4955	0.4956	0.4957	0.4959	0.4960	0.4961	0.4962	0.4963	0.4964
2.7	0.4965	0.4966	0.4967	0.4968	0.4969	0.4970	0.4971	0.4972	0.4973	0.4974
2.8	0.4974	0.4975	0.4976	0.4977	0.4977	0.4978	0.4979	0.4979	0.4980	0.4981
2.9	0.4981	0.4982	0.4982	0.4983	0.4984	0.4984	0.4985	0.4985	0.4986	0.4986
3.0	0.4987	0.4987	0.4987	0.4988	0.4988	0.4989	0.4989	0.4989	0.4990	0.4990

Using Table 7.1, Figure 7.4 shows which side of the distribution Z_1 and Z_2 (the standard deviation multipliers) are used. The entries in Table 7.1 are the fraction (percentage) of the total area under the curve measured from the mean of the distribution as the reference point to $Z_1\sigma$ or $Z_2\sigma$. We can, therefore, determine the percentage of population that falls between any two specified values using the table.

Referring to our numerical example, we may want to determine what percentage of items has a hole between the lower limit (LL) of 1.495 inches and the upper limit (UL) of 1.510 inches.

In order to do so, we relate these limit values to the population mean (1.502) and the standard deviation (0.003).

We can write two equations:

$$LL = \mu - Z_1 . \sigma$$
$$UL = \mu + Z_2 . \sigma$$

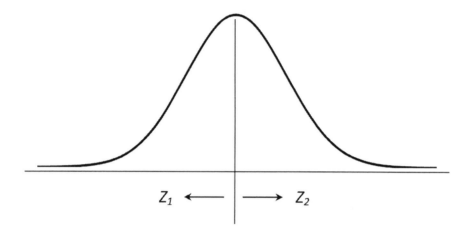

Figure 7.4. The essence of Z multipliers on a normal distribution curve.

Substituting numerical values:

$$1.495 = 1.502 - Z_1 . (0.003)$$
$$1.508 = 1.502 + Z_2 . (0.003)$$

Then, $Z_1 = 2.33$, and $Z_2 = 2.00$.

We note that Z_1 is to the left and Z_2 to the right of the mean. Now from Table 7.1 using $Z_1 = 2.33$, we find that 0.490 or 49.01% of the area is covered between the LL and μ, and using $Z_2 = 2.00$, we determine that 0.4772 or 47.72 percent of the area is covered between μ and the UL.** The sum of these two

---

** The corresponding table entry for a Z value is found as follows. For example, taking $Z_1 = 2.33$, we split it into a sum of two parts as: 2.30+ 0.03. The first term is located in the first column of the table. The second term is identified as column heading 0.03. The intersection of the row emanating for the first term and the column ema-

percentages, 96.73 percent, is the portion of the items that will have a hole diameter between 1.495 and 1.508 inches.

7.5 DISTRIBUTION OF SAMPLE AVERAGES

In some circumstances, the items in a population will have a normal distribution if they have been affected by several external factors. Similarly, cases can be found where the respective distribution is non-normal. It has been shown that if we:

- have any shaped distribution, containing a large population,
- take a sufficiently large number of samples of a specified size from this population,
- calculate the average of each sample,
- plot the frequency of occurrence for the averages thus obtained, the resulting distribution of the sample averages practically will have a normal distribution.

This distribution has a mean and a standard deviation, as does any other distribution. The statistical properties of this distribution are related to those of the main population when the number of samples taken approaches infinity or, for practical purposes, is at least very large.

If we show:

- Sample size by n
- Mean of the population by μ
- Average of each sample by $\bar{X}_i$
- Mean of the sample averages (grand average) by $\bar{\bar{X}}$
- Standard deviation of the population by σ
- Standard deviation of the sample means by $\sigma_{\bar{X}}$

Then in the long run:

$$\bar{\bar{X}} = \frac{\sum_{i=1}^{N} \bar{X}_i}{N} = \mu$$

$$\sigma_{\bar{X}} = \frac{\sigma}{\sqrt{n}}$$

nating from the second term as the column heading will read the value 0.4901. Knowing a value within the table, the corresponding Z value can be found by following the steps in reverse. The reason for this particular construct is to avoid a long two-column table.

$\sigma_{\bar{X}}$ is more commonly referred to as the *standard error of the sample averages.* $\bar{\bar{X}}$ and $\sigma_{\bar{X}}$ define the resulting normal distribution.

The transformation of a population into a normal distribution by sampling allows us to take advantage of the convenient and useful properties of the normal distribution.

7.6 STATISTICAL QUALITY CONTROL METHODS

There are various approaches and methods for different applications. We focus on the principal areas of application as depicted in Figure 7.5.

We now introduce these methods with respect to their areas of application and implementation in practice.

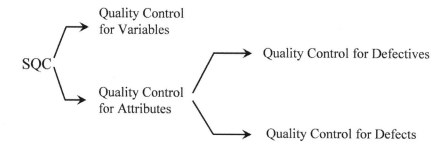

Figure 7.5. Statistical quality control for variables and attributes.

7.6.1 Quality Control for Variables: $\bar{X}$ and R Charts

There are products where the feature of interest, for example, a length, can be measured and expressed as a numerical value. In such cases the use of the method of "quality control for variables" is relevant. We argued that when a product is produced in large quantities, the measurements of a variable of interest will not all be the same despite the fact that a specific target value is ideally desired. A plot of the frequency of observations against the measurements will reveal the nature of the distribution of the measurements.

The properties of the normal distribution have been found to have a high utility in SQC. However, any variable so controlled must then be normally distributed. Fortunately, as outlined earlier, the "sampling" nature of SQC transforms any distribution, normal or non-normal, to a normal distribution of the sample averages allowing us to use the properties of normal distribution and its associated tables. Therefore, we will not be too concerned if the population being produced is not normal.

The quality of an item can be determined by measuring its desired feature and comparing it with the specifications for this feature. Specifications are typically given as a desired target value, commonly referred to as the mean value, plus/minus a tolerance. This tolerance need not be symmetric with respect to the mean value, though often it is. If the measurement of a produced part falls within the specification limits, the part is acceptable. Otherwise, it is considered as defective.

Assume that a manufacturer has received orders for a part whose critical feature is the depth of a drilled hole and the desired specification is given.

Accepting the orders is a sign that the firm believes it has the necessary combination of factors of production to produce such parts. In other words, since it is a fact that there will be variability in the depth of the holes produced, the firm will attempt to generate a population with a satisfactory mean and standard deviation such that an acceptable proportion of parts will fall within the specification limits.

In order to control the quality of this variable of interest in terms of sample averages, we need to set up a chart as shown in Figure 7.6.

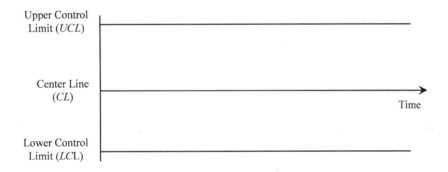

Figure 7.6. **Control chart template for sample averages.**

We now discuss the values of CL, LCL, and UCL in the chart. As outlined in Section 7.5, when we sample a distribution, we obtain $\mu = \overline{\overline{X}}$. We set the centerline $CL = \overline{\overline{X}}$. The values LCL and will UCL depend on how stringent the quality control must be. It has been experienced by industrial practitioners that a three-sigma limit satisfies the majority of quality control needs and, it is very common in practice. Using this concept, Figure 7.6 is shown again as Figure 7.7 in applying the control chart.

This chart is posted adjacent to the manufacturing equipment, and as items are produced, samples of a specified size are randomly taken from the output line, their average calculated and plotted on the chart. Inference is then made based on the behavior of the plotted points—or the so-called observations.

If the observations fall within the limits, we should only be partially happy. Since, as yet, we cannot rush to declare that the process is in control and acceptable output is being produced. Why? The reason is that whereas the sample averages may be fine, the "individual" items within a sample may have scattered values beyond the acceptable values. Therefore, the degree of spread or dispersion of the data must also be controlled. This may be done by calculating the standard deviation of each sample and controlling it in a similar fashion as the sample average. This, however, is rather cumbersome because we have to use a much larger sample size and then calculate the "standard deviation of the standard deviation of the samples." It has been shown that it is much more convenient and intuitive to work with the sample ranges. There exist statistical relationships between the standard deviation of the population being sampled and the data related to the range of the samples. Specifically, if we show:

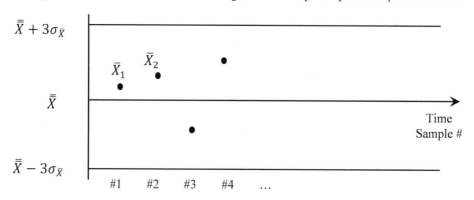

Figure 7.7. Control chart for sample averages.

- Sample size by n
- Standard deviation of the main population by σ
- Mean of the ranges obtained by $\bar{R}$
- Standard deviation of the sample ranges by σ_R

then:

$$\bar{R} = \frac{\sum_{i=1}^{N} R_i}{N}$$

and in the long run:

$$\bar{R} = d_2 \sigma$$

$$\sigma_R = d_3 \sigma$$

where d_2 and d_3 are obtained from a standard statistical data, Table 7.2.

σ_R is more commonly called the *standard error of the sample ranges.*

Table 7.2. Values of d_2 and d_3

Sample Size	d_2	d_3
2	1.128	0.835
3	1.693	0.888
4	2.059	0.880
5	2.326	0.864
6	2.534	0.848
7	2.704	0.833
8	2.847	0.820
9	2.970	0.808
10	3.078	0.797
11	3.137	0.787
12	3.258	0.778
13	3.336	0.770
14	3.407	0.762
15	3.472	0.755
16	3.532	0.449
17	3.588	0.743
18	3.640	0.738
19	3.689	0.733
20	3.735	0.729
21	3.778	0.724
22	3.819	0.720
23	3.858	0.716
24	3.895	0.712
25	3.931	0.709

If we construct a frequency plot of the sample ranges, we will have a distribution with an appearance as shown in Figure 7.8.

We see that the distribution of sample ranges is nearly normal. With the arguments we have made in this chapter that the normal distribution is convenient to use, we will treat this distribution as if it were normal. We can now set up another chart as shown in Figure 7.9.

This chart is posted next to the $\bar{X}$, chart, and as periodic samples are taken from the process output, the sample mean $\bar{X}$, and the sample range R are calculated and marked on the respective chart. The interpretation of the behavior of the observations is given later in Section 7.6.6.

Maintaining the sampled data within the limits of $\bar{X}$, and R charts is considered equivalent to maintaining μ and σ of the entire population at satisfactory levels.

We notice that $\bar{X}$, chart is based on μ, σ, and n, and R chart is based on σ, and n; n is needed to select the values of d_2 and d_3. But where do these values come from in the first place, especially, when a process is new and there is no recorded data?

The value for n is simply decided on, and there are guidelines for doing so. For μ and σ there are two approaches to determine their values: the "aimed-at-values" method, and the "estimated-values" method. We describe these next.

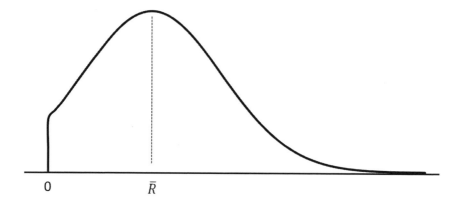

Figure 7.8. **Distribution of sample ranges.**

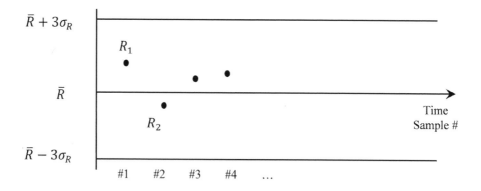

Figure 7.9. **Control chart for sample ranges.**

7.6.2 The Aimed-at-Values Method

This method is simple, straightforward, and requires only three elements:

1. An assumption that the main population being produced is normally distributed
2. A target value and tolerance limit for the variable of interest
3. An acceptable level of defectives is stated

With these, we can determine all that is needed to set up the $\bar{X}$, and R charts in a few short calculations.

The method is best demonstrated through an example. Let's assume that a population of parts is required with the specifications 4.624 ± 0.017 inches. It is reasonable to assume that the target value of the specification is the mean of the distribution of the part which will be generated. Thus, $\mu = 4.624$ inches. The standard deviation, σ, will depend on the tolerance values of ± 0.017 inches. To determine this standard deviation, we need to specify an acceptable level of defectives. For instance, if we consider this to be 2 percent, then 98 percent of the items must fall within the stated specification limits. As part of the procedure, we *assume* that the distribution of the measurements of the feature of interest is normal.

Using a plot of a normal distribution and Table 7.1, we match the numerical values as shown in Figure 7.10.

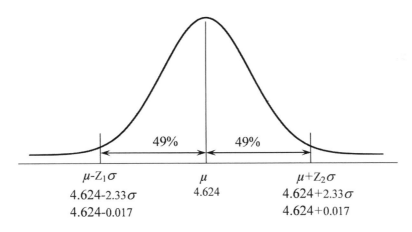

Figure 7.10. Matching data with normal distribution.

The 98 percent satisfactory output means that 49 percent of the output will be smaller, and 49 percent of the output will be larger than the mean. Using Table 7.1 and visually searching for the closest table entry to 49 percent (or 0.49), which is

0.4901, we find $Z_1 = Z_2 = 2.33$. Referring to Figure 7.10, the terms $4.624 \pm 2.33\sigma$ coincide with and are set equal to 4.624 ± 0.017, from which we obtain $\sigma = 0.0073$.

Now, using any reasonable value for the sample size, such as the common $n = 5$, and the determined values for μ and σ, we have all the information to proceed and establish $\bar{X}$, and R charts. Despite the great advantage of simplicity, we can point out a number of issues for this method. The first is the assumption of normal distribution of the population of parts, which may or may not be true. If the distribution is not normal, using the values determined to establish the control charts may lead to a percentage of defective other than 2 percent being produced. Second, the μ and σ determined are not unique. There can be other values, which with the same combination of factors of production, will still yield the desired population of parts. For these reasons the aim-at-values is not a favorable approach—though still a method of last resort. It can be used when there is evidence that the problems stated are minimal or not present.

7.6.3 The Estimated-Values Method

This method is more realistic, practical, and common. It relies on the process to be controlled itself rather than assumptions, but it requires some degree of effort and time. The procedure for this method is detailed below:

1. The process of interest is set up.
2. A reasonable and logical combination of factors of production is provided.
3. It is faithfully believed that the process will produce acceptable quality.
4. The process is commenced.
5. After a short settling period, twenty-five samples of a decided size are randomly taken from the output. Twenty-five is a rule of thumb and common in the industry. The sample size can be anywhere from two to ten or more, but five is common.
6. The data are recorded similar to the template shown in Table 7.3.

Table 7.3. Recoding sample data

Sample Number	Sample Average, $\bar{X}_i$ [unit]	Sample Range, R_i [unit]
1	...	...
2	...	...

3	...	...
.	.	.
.	.	.
.	.	.
25	...	...

7. $\overline{\overline{X}}$ and $\overline{R}$ are calculated as shown below.

$$\overline{X}_i = \frac{\sum_{j=1}^{n} X_j}{n}$$

$$\overline{\overline{X}} = \frac{\sum_{i=1}^{N} \overline{X}_i}{N}$$

$$\overline{R} = \frac{\sum_{i=1}^{N} R_i}{N}$$

Where n is the sample size, and N is the number of samples.

8. Using n, from Table 7.2 the values for d_2 and d_3 are selected.

9. From the relation $\overline{R} = d_2 \sigma$ we find σ.

10. From the relation $\sigma_R = d_3 \sigma$ we determine σ_R.

11. At this point, we have all the parameters needed to determine the control limits for the $\overline{X}$, and R charts. These limits, however, are referred to as the "trial control limits."

12. In order to verify the credibility and stability of the data, the twenty-five sample averages and sample ranges are checked against their respective chart. Knowing that we have used three-sigma limits (mean±3×sigma), then statistically in a normal distribution, only about three points in one thousand (that is, 0.3 percent) may fall outside the charts' limits. With twenty-five points we should not expect this to occur.

13. If all the sample averages and sample ranges fall within the limits of their respective chart, we drop the term *trial*, and call the control limits and the charts "final." If any of the points fall outside the chart limits, we cross out those points in the table of data, for instance, say, X_3, and R_6 and R_7, return to Step 7, and repeat the calculations with the remaining twenty-four sample averages and twenty-three sample ranges. The removed data points are called the outliers and are attributed to random variation, or to such factors as errors in measurement and insufficient operator learning period.

14. With the new values of $\bar{\bar{X}}$ and $\bar{R}$ determined in Step 7, we repeat Steps 9 to 13 until either we finalize the charts and go to Step 17, or too few sample data remain in Step 13 to make sound estimates, and we go to Step 15. The reason for the continued iteration is that the revised limits become narrower and shifted, which may and often do result in some of the remaining data points to fall outside the new limits.

15. If final charts cannot be established, we conclude that the factors of production need to be examined, and after fixing the suspected source(s) of the corrupt data, while having Step 16 in mind, we repeat the entire procedure from Step 1 and work with twenty-five fresh samples.

16. If multiple attempts to establish the $\bar{X}$, and R charts prove to be unsuccessful, we can say with good confidence that there are significant issues and the combination of the factors of production is inadequate for the process at hand. For example, we may have tight tolerances, but low precision machines to achieve them. Therefore, it is necessary to rectify the problem in a major way, such as acquiring more suitable equipment, lowering the expectation by relaxing the tolerances, or considering alternative designs and processes. This matter is dictated by the specific circumstances present.

17. If the charts are successfully established, periodic samples are taken, and $\bar{X}$, and R values are calculated and marked on the respective chart.

7.6.4 Evaluation of the Level of Control

After $\bar{X}$, and R charts have been established using the estimated-values method, the subsequent application may prove that the process behavior is acceptable, and it is in control. Although such may be the case, however, we should determine an indication of whether the estimated values of μ and σ will yield a satisfactory population of parts. At this point, the only option is to *assume* that the respective population distribution is normal, and by using the upper and lower specification limits estimate the percentages of acceptable and defective parts. We show this process graphically with the aid of Figure 7.11.

Suppose that the specification for the part remains as 4.624±0.017 inches as in our previous example, and by using the estimated-values method we have determined that $\mu = 4.6156$ inches, and $\sigma = 0.0053$ inches. Referring to Figure 7.11, we can write two equations:

At Lower Specification Limit (LSL): Target − Tol = $\mu − Z_1\sigma$
At Upper Specification Limit (USL): Target + Tol = $\mu + Z_2\sigma$

Substituting numerical values:

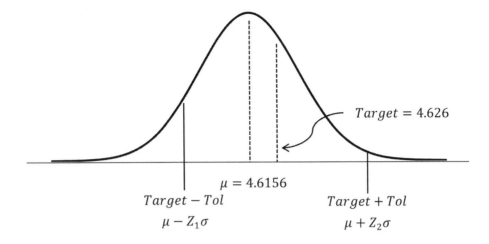

$Target = 4.626$

$\mu = 4.6156$

$Target - Tol$

$\mu - Z_1\sigma$

$Target + Tol$

$\mu + Z_2\sigma$

Figure 7.11. **Determining the level of control.**

$$4.624 - 0.017 = 4.6156 - Z_1 \times 0.0053$$
$$4.624 + 0.017 = 4.6156 + Z_2 \times 0.0053$$

From which we obtain $Z_1 = 1.62$, and from Table 7.1 we find that $0.5 - 0.4463 = 5.37\%$ of the parts will be undersize.

Similarly, we obtain $Z_2 = 4.79$, and since it is a large value, practically 0 percent of the parts will be oversize.

Therefore, with the assumption of normal distribution, 5.37 percent of the parts will be defective. Although this is an approximation, it cannot be too significantly in error, and a decision must now be made whether this is an acceptable level of defectives. If this is not the case, then corrective action must be taken. In this example, an attempt to increase the population mean toward the target specification will reduce the amount of defectives. This can be done by instructing the operator to make some adjustments when setting up the equipment used.

Example 7.1

A quality control analyst has decided to establish $\bar{X}$, and R Control charts at a machining center where a hole is drilled in a part. He instructed that thirty samples of size 4 should be taken from the process output, the diameter of the hole of each part contained in each sample measured, and the sample average and sample range calculated and recorded. The outcome is as shown in Table 7.4.

Table 7.4. Data for quality control example

Sample Number	Sample Average, $\overline{X}_i$ [in.]	Sample Range, R_i [in.]	Sample Number	Sample Average, $\overline{X}_i$ [in.]	Sample Range, R_i [in.]
1	1.0969	0.012	16	1.1008	0.012
2	1.1068	0.003	17	1.1034	0.003
3	1.1037	0.008	18	1.1040	0.005
4	1.1027	0.007	19	1.1080	0.008
5	1.0988	0.001	20	1.1061	0.002
6	1.1101	0.016	21	1.1050	0.011
7	1.0976	0.011	22	1.1068	0.009
8	1.0985	0.011	23	1.1074	0.013
9	1.0998	0.003	24	1.0984	0.032
10	1.1034	0.003	25	1.1030	0.006
11	1.0954	0.031	26	1.1040	0.007
12	1.1000	0.006	27	1.1001	0.005
13	1.1010	0.007	28	1.0960	0.007
14	1.0971	0.002	29	1.1052	0.012
15	1.0930	0.010	30	1.0999	0.003

a. Determine the values of the central line and the three-sigma limits for the control charts.

b. The current specification for the hole is 1.100 ± 0.008 inches. There is evidence that the individual measurements are normally distributed. Assuming the process remain in control at the present level; determine the percentage defective output expected from the process.

c. To improve the quality of the final product, the analyst has further decided to change the specification to 1 ± 0.007 inches. What percentage defective should the analyst expect?

Solution

(*a*)

We first obtain the values of the necessary parameter as follows:

$$\mu = \bar{\bar{X}} = \frac{\sum_{i=1}^{N} \bar{X}_i}{N} = \frac{33.0526}{30} = 1.1018 \; in.$$

$$\bar{R} = \frac{\sum_{i=1}^{N} R_i}{N} = \frac{0.2657}{30} = 0.00886 \; in.$$

These are the central lines for $\bar{X}$, and R charts, respectively. Using Table 7.2 with n = 4, we will also have $d_2 = 2.059$ and $d_3 = 0.880$. Then:

$$\bar{R} = d_2 . \sigma$$

$$\sigma = \frac{\bar{R}}{d_2} = \frac{0.00886}{2.059} = 0.0043 \; in.$$

$$\sigma_{\bar{X}} = \frac{\sigma}{\sqrt{n}} = \frac{0.0043}{\sqrt{4}} = 0.00215 \; in.$$

$$\sigma_R = d_3 . \sigma = 0.880 \times 0.0043 = 0.00378$$

We can now setup the *first trial control limits* for the $\bar{X}$, and R charts.

$$UCL = \bar{\bar{X}} + 3\sigma_{\bar{X}} = 1.1018 + 3 \times 0.00215 = 1.1082 \; in.$$

$$LCL = \bar{\bar{X}} - 3\sigma_{\bar{X}} = 1.1018 - 3 \times 0.00215 = 1.0953 \; in.$$

$$UCL = \bar{R} + 3\sigma_R = 0.00886 + 3 \times 0.00378 = 0.02021 \; in.$$

$$LCL = \bar{R} + 3\sigma_R = 0.00886 - 3 \times 0.00378 = -0.00249 \; in. < 0$$

Since the last value cannot be negative, it is set to zero. Thus:

$$LCL = 0.0$$

Next, as outlined in Section 7.6.3 under Step 12, we compare the values in Table 7.4 against the charts' limits and find possible outliers.

We notice that sample averages X_6 and X_{15} and the sample ranges R_{11} and R_{24} fall outside the control limits of their respective chart. We cross out these points in the table and recalculate $\bar{\bar{X}}$ and $\bar{R}$ again as follows:

$$\bar{\bar{X}} = \frac{\left(\sum_{i=1}^{N} \bar{X}_i \right) - \bar{X}_6 - \bar{X}_{15}}{N-2} = \frac{33.0526 - 1.1101 - 1.0930}{28} = 1.1020 \, in.$$

$$\bar{R} = \frac{\left(\sum_{i=1}^{N} R_i \right) - R_{11} - R_{24}}{N-2} = \frac{0.2657 - 0.031 - 0.032}{28} = 0.00723 \, in.$$

These are the new central lines for the $\bar{X}$, and R charts. We can repeat the previous calculations with these values. The parameters that remain unchanged are $d_2 = 2.059$ and $d_3 = 2.880$.

$$\bar{R} = d_2 . \sigma$$

$$\sigma = \frac{\bar{R}}{d_2} = \frac{0.00723}{2.059} = 0.00351 \; in.$$

$$\sigma_{\bar{X}} = \frac{\sigma}{\sqrt{n}} = \frac{0.00351}{\sqrt{4}} = 0.001755 \; in.$$

$$\sigma_R = d_3 . \sigma = 0.880 \times 0.00351 = 0.00309$$

We now set up the *second trial control limits* for the $\bar{X}$, and R charts:

$$UCL = \bar{\bar{X}} + 3\sigma_{\bar{X}} = 1.1020 + 3 \times 0.001755 = 1.10727 \; in.$$

$$LCL = \bar{\bar{X}} - 3\sigma_{\bar{X}} = 1.1020 - 3 \times 0.001755 = 1.0967 \; in.$$

$$UCL = \bar{R} + 3\sigma_R = 0.00732 + 3 \times 0.00309 = 0.01609 \; in.$$

$$LCL = \bar{R} - 3\sigma_R = 0.00732 - 3 \times 0.00309 = -0.00195 \; in. < 0$$

Since the last value is negative, it is set to zero. Thus:

$$LCL = 0.0$$

These are the new control limits for $\bar{X}$, and R charts. Checking the remaining twenty-eight data points against these control chart limits, we see that three new sample averages, X_{11}, X_{19}, and X_{28}, fall outside the control limits, but all the ranges are within their respective limits. We do have the option of removing both $\bar{X}$, and R values for the outlier samples; however, here we choose to remove only the $\bar{X}$, data. Then the control limits for the R chart are *final*. Similar to the procedure followed above, we cross out these points in the table of data, and proceed to determine yet another set of new control limits for the $\bar{X}$, chart. Since the R chart has been finalized, the only parameter that we must recalculate is the mean of the sample averages:

$$\bar{\bar{X}} = \frac{\left(\sum_{i=1}^{N} \bar{X}_i\right) - \bar{X}_6 - \bar{X}_{15} - \bar{X}_{11} - \bar{X}_{19} - \bar{X}_{28}}{N - 5}$$

$$= \frac{33.0526 - 1.1101 - 1.0930 - 1.0954 - 1.1080 - 1.0960}{28} = 1.1020 \; in.$$

This new value for the mean of the sample average happens to be the same as the previous value, and this is purely a coincidence. In this numerical example, the removed points have simply canceled their effect on the mean. Since no other parameters need to be calculated, we have the same mean for the averages, and with the outlier points having been removed, the previous limits determined in Trial 2 are retained and declared as final, and this precludes setting up the third trial control limits. For completeness, however, we restate the final control limits.

Control limits for the $\overline{X}$, chart:

$$UCL = \overline{\overline{X}} + 3\sigma_{\overline{X}} = 1.10727 \ in.$$
$$CL = 1.1020 \ in.$$
$$LCL = \overline{\overline{X}} - 3\sigma_{\overline{X}} = 1.0967 \ in.$$

Control limits for the R chart:

$$UCL = \overline{R} + 3\sigma_R = 0.01609 \ in.$$
$$CL = \overline{R} = 0.01609 \ in.$$
$$LCL = \overline{R} - 3\sigma_R = 0.00 \ in.$$

The charts will simply look as shown in Figure 7.12, which are posted adjacent to the process under study or made available electronically.

Solution

(b)

For Part (a) of the problem, since sampling was used, the assumption of normality and working with the respective equations were justified. In Part (b), we would like to determine what percentage of the main population of parts may be defective and outside the specification limits. For this purpose, we need to make an assumption that the population is normally distributed. This way we can certainly compute some numbers, but we must exercise caution as the underlying population may or may not be actually normal.

However, assuming a normally distributed population, and recalling that $\mu = \overline{X}$ from the given specification:

Lower Specification limit, LSL = 1.100 - 0.008 = 1.092 in., and

Upper Specification Limit, USL = 1.100+0.008 =1.108 in.

$$Z_1 = \frac{\mu - LSL}{\sigma} = \frac{1.1020 - 1.092}{0.00351} = 2.85$$

From Table 7.1 this yields the area under the curve as $A_1 = 0.4978$, and:

$$Z_2 = \frac{USL - \mu}{\sigma} = \frac{1.108 - 1.1020}{0.00351} = 1.71$$

From Table 7.1 this yields $A_2 = 0.4554$.

The portion of the area within the lower and upper specification limits is then $0.4978 + 0.4554 = 0.9532$.

The complement portion of the area outside the specification limits, $1 - 0.969 = 0.0468 = 4.68\%$ is defective.

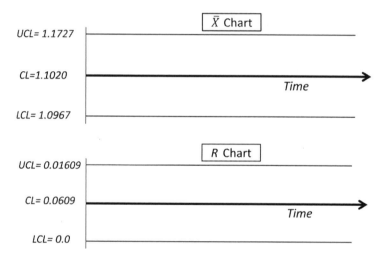

Figure 7.12. Final control charts for $\bar{X}$ and R.

Solution

(c)

We follow similar calculations as in Part (b) except that the specification used will be 1.100 ± 0.007 inches.

Lower Specification limit, LSL $= 1.100 - 0.007 = 1.093$ *in.*, and

Upper Specification Limit, USL $= 1.000 + 0.007 = 1.107$ *in.*

$$Z_1 = \frac{\mu - LSL}{\sigma} = \frac{1.1020 - 1.093}{0.00351} = 2.56$$

From Table 7.1 this yields the area under the curve as $A_1 = 0.4948$.

$$Z_2 = \frac{USL - \bar{\bar{X}}}{\sigma} = \frac{1.107 - 1.1020}{0.00351} = 1.42$$

From Table 7.1 again this yields $A_2 = 0.4222$.

The total area within the specification limits is $0.4948 + 0.4222 = 0.9170$.

The complement portion of the area outside the specification limits, $1 - 0.9170 = 0.083 = 8.30\%$ is defective.

Clearly, tightening the specification limits results in more parts being classified as defective and waste. On the other hand, the quality of the final product will be higher.

7.6.5 Distinction between Defective and Reworkable Parts

Quite often, a part that is classified as defective can be reworked into an acceptable one. An "external diameter" that is greater than the upper specification limit can be remachined to reduce its diameter, hopefully, to within the specification limits. Similarly, an "internal diameter" that is smaller than the lower specification limits can be rebored to within the specification limits. The "net" gain from recovering such a part must be compared with the loss value if the part were to be scrapped, and a decision made on an economic basis. If a part has successfully passed several other processes and already has a significant worth, then recovery cost and effort may be justified. This occurs for some consumer products, where the problems are corrected, and the products are sold with a label stating "factory refurbished."

7.6.6 Interpretation of Sample Behavior

Random chance variations are common and expected in many industrial processes. Using the established $\bar{X}$, and R charts, if an observation falls outside a control limit, a particular undesirable pattern is observed, or a trend is evident, investigation of the factors of production will often lead to discovering the source of the problem. Such a source is commonly referred to as an *assignable cause.* One of the main purposes of implementing a quality control program for a process is to locate such assignable causes and take appropriate steps to correct them. Learning lessons from industrial practice, below are some rules of thumb for the observations that call for an investigation of the factors of production while using the $\bar{X}$, and R charts:

- Seven consecutive points fall on one side of the center line.
- Ten out of eleven points fall on one side of the center line.

- Twelve out of fourteen points fall on one side of the center line.
- Fourteen out of seventeen points fall on one side of the center line.
- Sixteen out of twenty points fall on one side of the center line.

In day-to-day operations, we would ideally like the observations to follow natural and chance variations, and be fairly evenly spread about the center line. Other observations that also call for an investigation include the following:

- Seven consecutive points with upward or downward trend
- Points consistently too close to the control limits
- Points consistently too close to the centerline in $\bar{X}$, chart
- Points consistently too close to the lower control limit in R chart
- Points erratically moving from near one control limit to the other

A trend in one direction indicates the likelihood of future observations falling outside the control limit being approached. In the case of $\bar{X}$, chart, points too close to the center line, and in the case of R chart, points close to the lower limit can be interpreted as follows:

- The operator is spending excessive time to produce near perfect parts.
- The process has reached a stage that acceptable variations have become inherent.

These situations must be examined and adapted accordingly.

7.6.7 Nature and Frequency of Sampling

Care must be exercised to draw the samples from a closely related population. For instance, if two shifts or two operators are assigned to a particular process, it is advisable that separate charts are developed and used. Also, samples taken at the start of a shift may differ from the samples taken toward the end of the shift. The analyst must take these factors into consideration when making decisions about the behavior of the sampled data. The frequency of sampling will certainly depend on the rate of process output. At the beginning of implementing a quality control program, more frequent samples should be taken. Once it is ascertained that the process is in control, the frequency may be reduced over time.

7.6.8 Control Charts Limits

In our discussions and examples, we have mostly adhered to three-sigma limits. This originates from a host of practical industrial applications and experiences and is common in quality control. Note that it is not necessary to use an integer multiple of sigma. A quality control analyst may over time develop a deep insight into the process under study and determine that some other multiples of sigma, such as 2.7 or 3.1, work better and lead to a greater level of control in terms of the defectives generated. Let us now look at situations when other control limits are used. For example, if we use narrower 2.2-sigma limits, we will be too sensitive and reactive to natural variations and attempt to investigate the situation while no problems actually exist. This condition is called *Type I* error. Or if we use, for instance, wider 3.4-sigma limits, we may be too slow in attending to the evolving problems and miss the opportunity to rectify them. This condition is referred to as *Type II* error. A compromise that suits many practical situations is the common three-sigma limits. A quality control analyst, however, may decide to use what he or she argues to be more relevant limits for his or her application. For instance, in the aircraft industry, much narrower limits are used to maintain tight tolerances and precision and to detect developing problems quickly. Because of stringent manufacturing processes, the standard deviations are also generally smaller, leading to narrower control limits.

7.6.9 Areas of Application

In general, any process that generates a population with measurable features can be subjected to quality control for variables. Quality control efforts can cause a delay in operations and have cost implications. Therefore, only processes that are known or expected to produce a substantial amount of defective parts that must either be scrapped or reworked should be considered for quality control. Over time, the quality control personnel may find that it is adequate to maintain only one of the $\bar{X}$ and R charts, or abandon both charts altogether for a process when it remains steady and consistently produces satisfactory output. This occurs particularly with new processes that settle, and new operators who gain experience and improve their skills over time.

Another important factor that must be taken into consideration is the trade-off between the cost incurred by the quality control efforts and the cost of scraps. If the difference is insubstantial, it is advisable to implement a quality control program, because, in the long run, it will most likely prove to be a more economical choice.

7.7 QUALITY CONTROL FOR ATTRIBUTES

In Section 7.6.1, we stated that the quality control for variables required the quality of an individual item to be determined by first measuring the magnitude of a particular feature and then comparing it with that of the stated specification.

However, on many occasions, alternative approaches must be used. In some situations, the inspection process does not yield a quantitative measure of the part, but rather it indicates whether the part is defective or not. In such instances $\overline{X}$, and R charts could not be used. The quality control for attributes covers two distinct categories: quality control for *defectives* and quality control for *defects*. Respective methods are described next.

7.7.1 Quality Control for Defectives

There are many industrial situations where the inspection of the product or feature of interest does not yield a quantitative value, such as inspecting the quality of an electrical switch by subjecting it to an electrical test. The test will simply reveal whether the switch is faulty or not. Therefore, $\overline{X}$, and R charts are inapplicable for quality control in this case. In such circumstances quality control for defectives must be used. Other situations also arise, for example, when a go-no-go gauge is used to determine whether a particular dimension is within the stated specification. Again, $\overline{X}$, and R charts cannot be used. Go-no-go gauges are mechanical or electronic devices that are available for a variety of applications such as verifying inside diameters, outside diameters, lengths, and threads. A go-no-go gauge can quickly determine whether a part is within the desired specification. Figure 7.13 show a schematic of a mechanical go-no-go gauge for an outside diameter. A part is acceptable if it passes through the "go" gap, indicating that it is smaller than the upper specification limit,

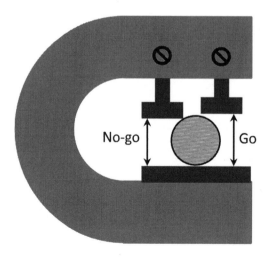

Go gap is set to Upper Specification limit and part must pass through it.

No-go gap is set to lower Specification limit and part must not pass through it.

Figure 7.13. Mechanical go-no-go gauge for the external diameter.

and does not pass through the "no-go" gap, indicating that it is larger than the lower specification limit. Otherwise, it will be classified as a defective or nonconforming part.

The circumstances leading to the application of the quality control for defectives can be recaptured as follows:

- When the inspection of the item of interest does not yield a quantitative measure.

- When quality control for variables is possible, but it is rather cumbersome, for example, when it is necessary to control the dimension of five different features of a part. This requires maintaining ten control charts. If the quality control for defectives is used only one chart, or if several features are combined into a small number of groups, much fewer charts are necessary.

- In situations where using control charts for variables is possible, but because of reasons such as the time required or the difficulty of inspection/measurement, quality control for defectives is preferred. Use of much faster go-no-go gauge instead of a micrometer is a good example.

In any of the above cases, the inspected part will simply be classified as good or defective. Therefore, detailed information will not be obtained, and this is a major disadvantage of using this method.

7.7.2 Number of Defectives in a Sample

The statistical concept to develop a control chart for defectives differs from the one on which the control charts for variables are based. The overall approach to quality control, however, is still the same.

A scenario that fits the application of quality control of a defective part is a population that is expected to contain mostly acceptable but also some defective or nonconforming parts. Such a population is expressed in terms of the percentage of the defective part, or what is commonly referred to as the proportion defective. As an example, if a population of ten thousand parts is produced, of which two thousand are defective, the proportion defective will simply be 0.20.

Now, suppose random samples of size 10 are taken from this population, and the items in each sample are inspected for quality and classified as acceptable or defective. Despite the fact that the population has a proportion defective of 0.2, we will find that each sample may and will have a different proportion defective. In other words, not every sample will contain exactly two defectives. Table 7.5 shows a possible outcome from taking thirty samples of size 10.

Table 7.5. Observation frequency of the number of defectives per sample

Number of Defectives per Sample	Sample Proportion	Number of Samples (Total:30)
0	0.0	xxx
1	0.1	xxxxxx
2	0.2	xxxxxxxxx
3	0.3	xxxxxx
4	0.4	xxx
5	0.5	xx
6	0.6	x
7	0.7	-
8	0.8	-
9	0.9	-
10	1.0	-

We notice that each observation appears with different frequency. The most likely number of defectives per sample is two, and the most likely sample proportion is 0.2. For a large number of samples, the plot of the relative frequency will be as shown in Figure 7.14. This frequency plot is called the "binomial distribution."

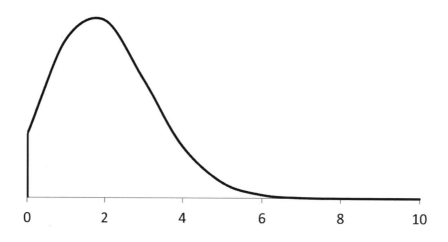

Figure 7.14. Binomial distribution of the number of defectives per sample.

Identically, Figure 7.14 can also be expressed in terms of the sample proportions as shown in Figure 7.15.

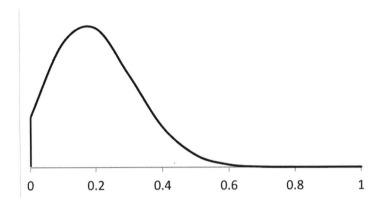

Figure 7.15. **Binomial distribution of the proportion defective per sample.**

Binomial distribution has a mathematical expression from which the probability of the occurrence of a particular range of values can be found. In statistical quality control and for the reasons that will be explained in the next section, we do not resort to computing such probabilities.

7.7.3 Mean and Dispersion

As with any other distribution, binomial distribution has a mean and dispersion. If we work with sample proportions, the mean of the distribution is the average of all sample proportions, and we show this by μ_p. It has been found that as the number of samples taken approaches infinity, this mean will be equal to the proportion defective of the entire population, p. That is:

$$\mu_p = \frac{\sum_{i=1}^{N} p_i}{N} = p$$

The standard deviation can be calculated using the conventional standard deviation formula. However, it has been shown that when the number of samples taken approaches infinity, a much simpler expression can be used:

$$\sigma_p = \sqrt{\frac{p(1-p)}{n}}$$

This is commonly called "Standard Error of the Proportions." As expected, this equation shows that as n increases, σ_p decreases.

Considering the distribution of the number of defectives per sample, Figure 7.14, similar equations can be obtained for its mean and dispersion. If p_i represents the proportion of defectives for a sample, then:

$$\text{Number of defectives in the sample} = np_i$$

The mean of the distribution, μ_{np}, will be the average of all the "number of defectives per sample." The number of defectives per sample is also called the "occurrences." When the number of samples approaches infinity, the mean of the distribution can be determined as:

$$\mu_{np} = \frac{\sum_{i=1}^{N} np_i}{N} = n \frac{\sum_{i-1}^{N} p_i}{N} = n\mu_p = np$$

where N is the number of samples. The standard error of the occurrences can be determined as follows:

$$\sigma_{np} = n\sigma_p = n \sqrt{\frac{p(1-p)}{n}} = \sqrt{np(1-P)}$$

It is clear from Figures 7.14 and 7.15 that binomial distribution is unsymmetric and does not extend from minus infinity to plus infinity. However, for statistical quality control purposes and the ease of use in practice, we treat the binomial distribution is if it were a normal distribution.

The justification for overlooking the errors induced is that in statistical quality control, the control limits of the charts are important; however:

1. We use three-sigma control limits invariably for many processes. There will always be some inherent approximation. Over time the quality control personnel from their observation of the process behavior and the rate of defective output will eventually fine-tune the process and the limits and, therefore, reduce the errors resulting from all approximations.

2. It is not solely the precise control limits that help maintain a process in control. In most circumstances, it is the behavior and the trend of the observations within the control limits that are used to detect issues and rectify the factors of production in an attempt to control the quality of future output.

7.7.4 Quality Control for Defectives: *p*-Chart and *np*-Chart

Let us consider how quality control charts can be set up for defectives. In the quality control for variables, we had to set up two charts, $\bar{X}$, chart for sample averages and *R* chart for sample ranges. These two charts were independent from each other. In the case of quality control for defectives, let us restate the parameters involved:

$$\mu_p = p$$

$$\sigma_p = \sqrt{\frac{p\,(1-p)}{n}}$$

We see that with *n* being constant, σ_p depends strictly on *p*. If we control *p*, we will also implicitly control σ_p. Therefore, we only need to maintain one chart for *p*. Assuming normal distribution and the common three-sigma limits a control chart for "proportion defective" or *p*-chart will be as shown in Figure 7.16.

In a similar fashion, using μ_{np} and σ_{np}, we can construct an *np*-chart for the "number of defectives per sample." This is shown in Figure 7.17.

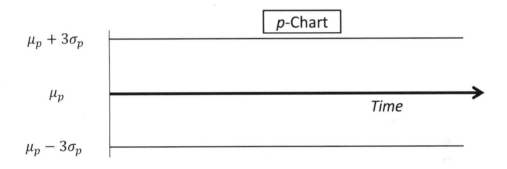

Figure 7.16. The *p*-chart for proportion defective.

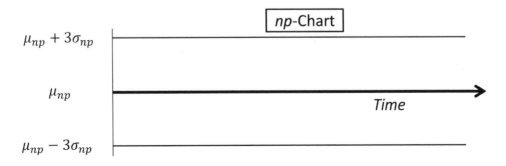

Figure 7.17. The *np*-chart for number of defectives per sample.

Both charts serve exactly the same purpose. The difference is whether it is preferred to work with proportion defective or the number of defectives per sample.

We notice that both the p-chart and the np-chart are based on p and n. But how do we obtain the values of p and n in the first place? For n, we wait until we discuss the subject in Section 7.7.7. There are two ways to determine p: the "aimed-at-value" and the "estimated-value" approaches. In the aimed-at-value method, the population proportion is assumed to be the desired or target value of the proportion. Such a value may or may not be actually true, and in the case that it isn't, it can lead to improper control limits and the wrong decision being made. For this reason, it is more desirable to use the estimated-values method, as described next.

7.7.5 Estimated-Value Method

As with the quality control for variables, in this case, too, the approach relies on the process being controlled itself but requires some degree of effort and time. The procedure for this method is detailed below:

1. The process of interest is set up.
2. A reasonable and rational combination of factors of production is provided.
3. It is believed that the process will yield an acceptable population.
4. The process commences.
5. After a short settling period, twenty-five samples of a decided size are taken. Twenty-five is again a rule of thumb and common in the industry. For the sample size, see Section 7.7.7.
6. A record of data is arranged as shown in Table 7.6.

Table 7.6. **Recording sample data**

Sample Number	Number of Defectives per Sample, np_i	Sample Proportion, p_i
1	...	...
2	...	...
3	...	...
.	.	.
.	.	.
.	.	.
25	...	...

7, μ_p, p and σ_p are calculated as shown below:

$$p = \mu_p = \frac{\sum_{i=1}^{N} p_i}{N}$$

$$\sigma_p = \sqrt{\frac{p\,(1-p)}{n}}$$

8. At this point, we have all that is needed to determine the control limits for the p-chart. These limits, however, are referred to as the *trial control limits.*

9. In order to verify the credibility and stability of the data, the twenty-five sample proportions are checked against the control limits. Knowing that we have used three-sigma limits, only about three points in one thousand (or 0.3 percent) are expected to fall outside the control chart limits.

10. If all the points are within the control limits, we drop the term *trial*, call the control limits and the chart "final," and go to Step 14. If any of the sample proportions fall outside the chart limits, we cross out those samples in the table of data, return to Step 7, and repeat the calculations with the remaining sample data.

11. With the new p (or identically μ_p) we repeat Steps 8 to 11 until either we finalize the chart and go to Step 14, or too few sample data points are left to make a sound estimate of p.

12. If the final chart cannot be established, we conclude that the factors of production need to be reexamined, and after fixing the suspected sources of the corrupt data, we take twenty-five fresh samples and repeat the entire procedure.

13. If attempts to establish a p-chart prove unsuccessful, we can say with confidence that the factors of production are unsuitable for achieving the expected output. At this point, a decision is made to rectify the problem, for instance, by acquiring suitable equipment, lowering the expectations, or modifying the design. The exact nature of any remedy will depend on the case at hand.

14. If the chart is successfully established, periodic samples are taken and the proportion defectives are calculated and marked on the chart.

Example 7.2

A quality control engineer has taken twenty samples of size 200 from the output of an assembly line. The items in each sample have been inspected, and the number of defectives in each sample recorded. The results are given in Table 7.7.

Table 7.7. Data for quality control example

Sample Number	Number of Defectives, np_i	Sample Number	Number of Defectives, np_i
1	18	11	12
2	11	12	13
3	7	13	10
4	12	14	11
5	16	15	9
6	7	16	9
7	14	17	10
8	15	18	15
9	9	19	8
10	26	20	8

a. Basing your estimate of the population mean on these results, compute the values of the central line and two-sigma control limits for a p-chart that the engineer wants to maintain at the assembly line.

b. Once the control chart has been in use for a while, the engineer takes a sample of three hundred items and determines that it contains twenty-five defectives. Can the engineer assume that a satisfactory output is being produced?

Solution

(a)

It is equally applicable to set up either p or np chart. Data are given in terms of np, but a p-chart is required in this example. By dividing the sample data in Table 7.7 by the sample size, we obtain the data in terms of the proportions as shown in Table 7.8.

Table 7.8. Data expressed in proportion defective

Sample Number	Proportion Defective, p_i	Sample Number	Proportion Defective, p_i
1	0.090	11	0.060
2	0.055	12	0.065
3	0.035	13	0.050

4	0.060	14	0.055
5	0.080	15	0.045
6	0.035	16	0.045
7	0.070	17	0.050
8	0.075	18	0.075
9	0.045	19	0.040
10	0.130	20	0.040

We calculate the required measures as follows:

$$\mu_p = \frac{\sum_{i=1}^{N} p_i}{N} = \frac{1.2}{20} = 0.06 = p$$

$$\sigma_p = \sqrt{\frac{p(1-p)}{n}} = \sqrt{\frac{0.06(1-0.06)}{200}} = 0.0168$$

Next we determine the control limits.
First trial limits:

$$p + 2\sigma_p = 0.06 + 2 \times 0.0168 = 0.0936$$

$$p - 2\sigma_p = 0.06 - 2 \times 0.0168 = 0.0264$$

Checking the data in Table 7.8 against these limits, we find that $p_{10} = 0.13$ falls outside. We cross out this outlier and recompute the parameters:

$$\mu_p = \frac{\sum_{i=1}^{N} p_i - 0.13}{N-1} = \frac{1.2 - 0.13}{19} = 0.0563 = p$$

$$\sigma_p = \sqrt{\frac{p(1-p)}{n}} = \sqrt{\frac{0.0563(1-0.0563)}{200}} = 0.0163$$

Second trial limits:

$$p + 2\sigma_p = 0.0563 + 2 \times 0.0163 = 0.0889$$

$$p - 2\sigma_p = 0.0563 - 2 \times 0.0163 = 0.0237$$

We see that in this round $p_1 = 0.09$ falls outside the limits. We cross out this outlier and recompute the parameters for a third time:

$$\mu_p = \frac{\sum_{i=1}^{N} P_i - 0.13 - 0.09}{N - 2} = \frac{1.2 - 0.13 - 0.09}{18} = 0.0544 = p$$

$$\sigma_p = \sqrt{\frac{p(1-p)}{n}} = \sqrt{\frac{0.0544(1-0.0544)}{200}} = 0.0160$$

Third trial limits:

$$p + 2\sigma_p = 0.0544 + 2 \times 0.0160 = 0.0864$$

$$p - 2\sigma_p = 0.0544 - 2 \times 0.0160 = 0.0224$$

An examination of the data reveals that all fall within the above limits. These limits are final. The final p-chart for the assembly line is shown in Figure 7.18.

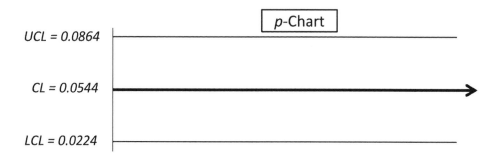

Figure 7.18. The p-chart for the assembly line.

Solution

(*b*)

With the information given, we compute:

$$p_j = \frac{25}{300} = 0.0833$$

We know that the larger the sample size, the closer should be its proportion to the population proportion. Clearly, this is not the case, and this sample proportion is very close to the upper control limit of the determined p-chart. The engineer ought to be concerned about the value 0.0833 and investigate, and, thus, he or she cannot make a credible judgment on the basis of one observation whether the process is in control or not.

Some Observations about the Preceding Example

We see that the analyst has taken only twenty random samples rather than the common practice of taking twenty-five samples. We also notice that the analyst has adhered to a two-sigma limit. Once a control chart has been established using a particular sample size, it is advisable that the size of the subsequent samples remain the same. Here, instead of confining to a sample size of two hundred, the analyst has taken a sample of three hundred items. From these efforts, we can say that the analyst is perhaps exploring the options to arrive at some optimal quality control conditions.

7.7.6 Interpretation of Sample Behavior

The same rules of thumb as outlined in Section 7.6.6 for $\overline{X}$, and R charts are also used for the p-chart.

7.7.7 Sample Size

The larger the sample size, the more likely it is that the sample proportion will approach the population proportion. Although this reduces the risk of making a wrong decision, the cost of sampling will be higher. A common rule of thumb is that both np and $n(1 - p)$ are equal or greater than 5. For most manufacturing processes and products, the proportion p is much smaller than its complement $1 - p$. Therefore, np is the deciding factor. However, n and p are both generally unknown, unless reliable historical data exits for p. To circumvent this uncertainty, a simpler rule is that the sample should be greater than one hundred. Gaging the process at hand, the analyst can exercise some thoughtful judgment and decide on the sample size. Once p has been estimated through sampling and more data have become available, the sample size can be better estimated and the control limits revised.

7.8 CONTROL CHART FOR DEFECTS: c-CHART

If a product is such that it can contain an infinite number of defects, then the control chart for defects or c-chart is used. It may seem rather frightful as to how an infinite number of defects in a sample is possible. This is only a modeling analogy, and we can identify real cases where we can see the relevance of this concept. Let us consider a sheet of glass. We can say that the sheet of glass is a single unit, a sample of size 1. We can also treat it as a container or *a sample of infinite size* because we can subject it to a countless number of scratches or defects, and it can accommodate them. In practice, however, we will neither speak this way nor deal with a large number of defects. Normally, we inspect the sheet of glass or similar product as a single item, hoping for zero, but possibly expecting only a small and finite number of defects—such as one, three, or maybe even twenty. With such a number of defects and a conceptual sample size of infinity, the proportion of defects will be close to zero. If a large number of such units of products are produced, and the number defects contained in each is counted, a plot of the observation frequency can be obtained. The respective population constitutes a special case of the binomial distribution called the Poisson distribution. The major difference between the binomial and Poisson distributions is that unlike binomial distribution, Poisson distribution is much more skewed to the right and extends to infinity in that direction.

In developing a control chart for defectives, we showed that the number of defectives per sample is explained by the binomial distribution with:

$$\mu_{np} = \frac{\sum_{i=1}^{N} np_i}{N} = n\frac{\sum_{i=1}^{N} p_i}{N} = n\mu_p = np$$

$$\sigma_{np} = n\sigma_p = n\sqrt{\frac{p(1-p)}{n}} = \sqrt{np(1-P)}$$

Since np represents the "number of defectives in a sample," it is analogous to expressing the "number of defects in a unit-sample," c. We can write, therefore, for the mean:

$$\mu_c = c$$

p is also analogous to the fraction defectives (here fraction defects) in a conceptual sample of infinite size. Therefore, it represents a proportion such that $p \ll 1$, and thus $1 - p \approx 1$. Then the above expression for the dispersion can be tailored for the number of defects into a very simple form:

$$\sigma_{np} = \sqrt{np(1-p)}$$
$$\sigma_{np} = \sqrt{np}$$
$$\sigma_c = \sqrt{c}$$

Figure 7.19 shows a Poisson distribution.

Having determined the mean and the standard deviation, everything about setting up a control chart for defects, referred to as the c-chart, from estimating the initial value of c to interpreting the observations, is similar to those of the $\overline{X}$, R and p charts.

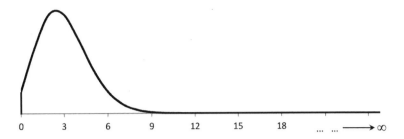

Figure 7.19. **Poisson distribution.**

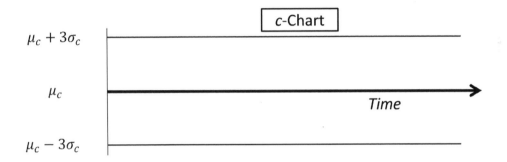

Figure 7.20. **Blank *c*-chart.**

A typical blank *c*-chart is shown in Figure 7.20.

7.8.1 Counting Defects

In quality control for defects, we gave an example of a sheet of glass that theatrically could contain an infinite number of defects. This argument was made to signify that the sheet of glass could be a sample of infinite size. In practice, we deal with very few defects per unit sample. We must, however, also quantify the severity of the observed defects. A corner-chip is significantly different than, say, a long scratch in the middle of the sheet of glass. The corner-chip can be readily edge-trimmed, leaving a smaller, defect-free part, whereas a long scratch can render the sheet of glass useless and of little value. For this reason, defects can be given weights or point values. A typical classification is shown in Table 7.9.

Table 7.9. Classification of defects

Class of Defect	Point Value
Very serious	3.0
Serious	2.0
Moderately serious	1.0
Not serious	0.5

Once point values have been assigned to all the identified defects, the sum of the point values is considered as the *effective number of defects* in the unit sample.

7.9 REVISING THE CONTROL LIMITS

To establish $\bar{X}$, R, p, and c charts, we assumed that we are dealing with a process that is new, or suitable prior data are unavailable. Even if it were, with the intent to implement a quality control scheme, we first tighten up the process by ascertaining that an appropriate combination of factors of production is provided and used. This will generate a new population that may differ from an existing one. Therefore, it is always better to use fresh samples from the fine-tuned process. To expedite the implementation of a quality control program, it is suggested that we begin with only twenty-five samples. However, once the respective charts have been successfully established and indicate that the process is producing an acceptable output, the control limits of the charts must be revised to better reflect the settling and stabilization of the process. This is done by taking, say, twenty-five additional samples, and using the fifty sample data available for reestimating the values of the required parameters. If these values significantly differ from the original estimates, the control limits can be revised. Revising the control limits should be periodically done by taking more samples. Older samples can be discarded. It is reasonable to maintain one hundred to two hundred most recent sample data on an ongoing basis.

EXERCISES

7-1 What is considered bad quality?

7-2 What are the consequences of bad quality?

7-3 Should every process have quality control in place?

7-4 Who decides on the level of tolerances and quality?

7-5   How would one obtain a normal distribution from a distribution that is not normal? How are the two distributions related?

7-6   To produce products of good quality and gain a reputation for it, a new company has decided to start off with a two-sigma limit on its control charts for variables. For a particular part whose critical dimension is its thickness, the company's quality control analyst begins by taking thirty samples of size 8, which he plans to use to estimate the population mean and standard deviation and set up the control charts. The analyst finds that the sum of the samples average is 7.822 inches and that the sum of the sample ranges is 0.987 inch.

  a.   If the process is assumed in control, what are the values of the central line and control limits for $\overline{X}$, and R charts?

  b.   If checking the observations against the charts results in a satisfactory assessment, the process continues to remain in control, and the population being generated is reasonably normally distributed; and the specification is 0.300 ± 0.020 inches, what percent of the output will be defective ?

  c.   Should oversize items always be classed as defective?

  d.   The company has been wise to start off with two-sigma limits. What else did the company do right?

7-7   A quality control analyst takes twenty-four samples of size 200 each from the output of a production line. The items in each sample are tested, and the number of defective items in each sample is recorded as shown in Table 7.10.

Table 7.10. Number of defectives in the samples

Sample Number	Number of Defectives	Sample Number	Number of Defectives
1	8	13	12
2	13	14	14
3	14	15	12
4	10	16	10
5	27	17	9
6	17	18	9
7	11	19	10
8	8	20	16

9	17
10	12
11	13
12	15

21	7
22	22
23	18
24	8

Determine the values of the central line and two-sigma control limits for a chart for fraction defective (p-chart) that the analyst must maintain at the production line.

7-8 A manufacturer of outdoor digital advertising display panels is planning to implement a quality control scheme. The displays are fifty LEDs wide and two hundred LEDs long. It has been observed that some LEDs on the panels fail to function. Such an LED is considered a defect in the product. (Although it does not affect your calculations, the company counts two adjacent flaws as three defects.) The average number of defects per unit of output is estimated to be ten and is considered satisfactory. For quality control purposes, the quality control analyst treats the actual distribution of defects per sample as if it were a Poisson distribution; $50 \times 200 = 10,000$ is large enough to assume a sample of size infinity:

a. What should be the values for the central line and two-sigma control limits for a c-chart?

b. Is two-sigma control limits a reasonable choice for the type of product? Explain your opinion.

Chapter 8 FACILITY LOCATION

8.1 INTRODUCTION

The increase in demand for a product or service as well as upgrading, renovation, and expansion of business require procurement, installation, or construction of new equipment and facilities. Typically, the new equipment or facility will have significant interaction in terms of material or vehicular movement with the existing facilities. Transportation costs are among the largest components of expenditure in the operation of many enterprises. Therefore, to minimize the costs, it is essential to determine an optimal location for the new facility to be established. There are numerous examples of where a new facility may be needed, such as the following:

- Warehouse to provide supplies for a number of outlets
- Tool dispenser in a manufacturing plant
- Library, fire station, or police station in a residential district
- Suburban airport to provide services for remote communities
- Power-generating plant
- Boiler in a processing plant

In addition to minimizing the direct transportation costs, there are instances where some other measures of optimality could be of interest. The response of a fire truck to fire is not primarily driven by the cost, but rather the travel time or distance. Therefore, for such cases, the goal is to minimize the maximum distance to travel. In this chapter, we review both cost and distance minimization for location problems.

Location problems can also be categorized into two classes:

1. Single new facility
2. Multiple new facility

We focus on the single-facility location problems.

8.2 FORMS OF DISTANCE

The common concept of distance is the straight line between two points. This is true as a strict geometric definition and can be used in many location problems, such as in an airport where aircraft take off and fly much of the distance to their destination in a straight line. However, there are other instances where different concepts of distance must be used. In a factory environment, the machinery and facilities are usually laid out in such a way that movement or transportation is accomplished through perpendicular aisles and passages. This form of movement and distance is referred to as the rectilinear or rectangular.

In one class of problems, the cost is proportional to the square of the "straight-line" distance. In some rare problems, which are outside the scope of this book, the cost may be proportional to an unconventional definition of distance, such as hyper-rectilinear. It is customary to use the positive quadrant of a Cartesian coordinate system to express distances and places for the facility location problems.

8.3 OBJECTIVE FUNCTION FORMULATION

First, we explain the formulation of the objective function for two classes of new facility location problems:

• **Minimization of total cost—mini-sum problems**

We consider m existing point-location facilities identified as $P_1, P_2, \dots P_m$, where a new facility is to be located at a point $Q(x,y)$, given that transportation costs are incurred and are proportional to an applicable form of distance between the new and the existing facility P_i.

If $d(Q, P_i)$ represents the distance traveled per trip between the points Q and P_i; and if w_i represents the product of cost per unit distance traveled and the number of trips per unit time between the new facility and the existing facility i, then the total travel cost per unit time (typically a year) is given by:

$$f(Q) = f(x, y) = \sum_{i=1}^{m} w_i d(Q, P_i)$$

The cost factors w_i are referred to as the "weights." The objective is to determine the coordinates of the new facility, Q^*, such that the total cost per unit time is minimized. The typical units for the terms in the equation are as follows:

$f(x, y)$ [dollar/unit time]

w_i [dollar/distance] [trips/unit time]

$d(Q, P_i)$ [distance/trip]

- **Minimization of maximum distance—mini-max problems**

In this class of problems, no direct cost is involved, and the objective function is stated as:

$$Minimize\ f(Q) = f(x, y) = \max d_i(Q, P_i), \quad i = 1, m$$

Notice that the formulation does not involve any form of weights.

8.4 MINI-SUM PROBLEMS

We consider three mini-sum new facility location problems where the transportation and movement costs are proportional to the following:

- Squared Euclidean distance
- Euclidean distance
- Rectilinear distance

Figure 8.1 shows the geometry of Euclidean and rectilinear distances.

For each case, we also present a method of determining near-optimal alternative locations when the optimum location found is not available or feasible.

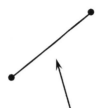

Euclidean distance
between two points.
One single path.

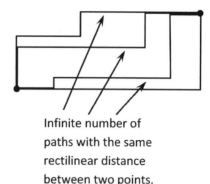

Infinite number of
paths with the same
rectilinear distance
between two points.

Figure 8.1. Euclidean and rectilinear distances.

8.4.1 Squared Euclidian Distance

The "Euclidean distance" is the measure of a straight line between two points. In some cases, we encounter Squared-Euclidean distance problems when the cost is proportional to the square of the Euclidean distance moved or traveled. An example of an application that fits this definition is the heat loss in power plant steam transmission lines. The Squared-Euclidean has a simple and straightforward solution, and it lays the ground for a better understanding of the other forms of location problems.

If the coordinates of the new facility Q are (x, y), and for an existing facility P_i, are (a_i, b_i), the Euclidean distance between the two points is defined as:

$$d_i(Q, P_i) = d_i = \left[(x - a_i)^2 + (y - b_i)^2\right]^{1/2}$$

The Squared-Euclidean distance is then:

$$d_i^2 = \left[(x - a_i)^2 + (y - b_i)^2\right]$$

The Squared-Euclidean problem may now be stated as:

$$\text{Minimize } f(Q) = f(x, y) = \sum_{i=1}^{m} w_i d_i^2$$

or

$$\text{Minimize } f(Q) = f(x, y) = \sum_{i=1}^{m} w_i \left[(x - a_i)^2 + (y - b_i)^2\right]$$

The typical units for the terms used in the above equation are:

$f(x,y)$ [dollar/unit time]

w_i[dollar/distance-square][trips/unit time]

$d_i(Q, P_i)$[distance/trip]

The approach to determining the coordinates of the new facility is to compute the partial derivatives of the cost function with respect to x and y, set them to zero, and solve:

$$\frac{\partial(x, y)}{\partial x} = 0$$

$$\frac{\partial(x, y)}{\partial y} = 0$$

or

$$\frac{\partial(x, y)}{\partial x} = \sum_{i=1}^{m} 2w_i \left(x - a_i\right) = 0$$

$$\frac{\partial(x, y)}{\partial y} = \sum_{i=1}^{m} 2w_i \left(y - b_i\right) = 0$$

Leading to:

$$x^* = \frac{\sum_{i=1}^{m} w_i a_i}{\sum_{i=1}^{m} w_i}$$

$$y^* = \frac{\sum_{i=1}^{m} w_i b_i}{\sum_{i=1}^{m} w_i}$$

The point $Q^* = (x^*, y^*)$ is the optimum location that minimizes the total cost of interaction per unit time used.

Owing to resemblance to the common weighted average calculations, such as the grade point average (GPA) in academic transcripts, the Squared-Euclidean solution is also referred to as the "centroid" or the "center of gravity" solution.

Iso-Cost Contour Lines

There is always the possibility that the optimum location determined is not readily available because of inaccessibility and physical limitations, such as a body of water or a highway. In such cases, it would be desirable and indeed helpful to construct constant- or iso-cost contour lines where a multitude of feasible alternative locations can be found. To demonstrate the concept of contour lines, we begin with two simple cases. When there is only one existing facility, the optimum new facility location coincides with it, and regardless of its weight, the iso-cost contour lines are concentric circles centered on the optimum location, as shown in Figure 8.2(a). If there are two existing facilities with equal weights, the optimum point is halfway between the two existing facilities and the iso-cost contour lines are also concentric circles, and centered on the optimum point, as shown in Figure 8.2(b).

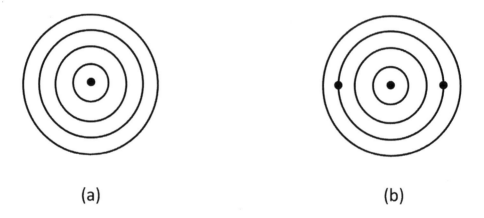

(a)   **(b)**

Figure 8.2. **Contour lines for simple Squared-Euclidean distance problems.**

Interestingly, the iso-cost contour for any number of existing facilities, even with unequal weights, will still be concentric circles centered on the optimum point. We derive this mathematically because it is useful when searching for an alternative near-optimal new facility location.

For the optimum location obtained, the least cost is:

$$C_{min} = f\left(Q^*\right) = f\left(x^*, y^*\right) = \sum_{i=1}^{m} w_i \left[\left(x^* - a_i\right)^2 + \left(y^* - b_i\right)^2\right]$$

For any $C_c > C_{min}$, where C_c is an acceptable cost, the set of all points that satisfy the equation:

$$C_C = f\left(x,y\right) = \sum_{i=1}^{m} w_i \left[\left(x - a_i \right)^2 + \left(y - b_i \right)^2 \right]$$

yields the contour line for a fixed cost of C_c.

Expanding and factorizing this equation:

$$C_C = x^2 \sum_{i=1}^{m} w_i - 2x \sum_{i=1}^{m} w_i a_i + \sum_{i=1}^{m} w_i a_i^2 + y^2 \sum_{i=1}^{m} w_i - 2y \sum_{i=1}^{m} w_i b_i + \sum_{i=1}^{m} w_i b_i^2$$

Let:

$$W = \sum_{i=1}^{m} w_i$$

Dividing both sides of the cost equation by W, we obtain:

$$\frac{C_C}{W} = x^2 - 2x \sum_{i=1}^{m} \frac{w_i a_i}{W} + \sum_{i=1}^{m} \frac{w_i a_i^2}{W} + y^2 - 2y \sum_{i=1}^{m} \frac{w_i b_i}{W} + \sum_{i=1}^{m} \frac{w_i b_i^2}{W}$$

Recalling and substituting $\dfrac{\sum_{i=1}^{m} w_i a_i}{\sum_{i=1}^{m} w_i} = x^*$ and $\dfrac{\sum_{i=1}^{m} w_i b_i}{\sum_{i=1}^{m} w_i} = y^*$ we have

$$\frac{C_C}{W} = x^2 - 2xx^* + \sum_{i=1}^{m} \frac{w_i a_i^2}{W} + y^2 - 2yy^* + \sum_{i=1}^{m} \frac{w_i b_i^2}{W}$$

Adding constant terms $(x^*)^2$ and $(y^*)^2$ to both sides of the above equation, factorizing and simplifying we obtain:

$$\left(x - x^* \right)^2 + \left(y - y^* \right)^2 = r^2$$

This is the equation for a circle where the radius is:

$$r = \left[\frac{C_C}{W} + \left(x^* \right)^2 + \left(y^* \right)^2 - \sum_{i=1}^{m} \frac{w_i \left(a_i^2 + b_i^2 \right)}{W} \right]^{1/2}$$

Alternative new facility locations can be found for various values of C_c, where increasing C_c will result in a larger radius, and clearly constitute points farther away from the optimum location. It is a combination of cost, feasibility, and perhaps some nonquantitative consideration that will help select the new location.

Example 8.1

A new facility is to be located with respect to four existing facilities. The interaction between the new facility and the existing facilities is such that the cost is proportional to the squared Euclidean distance. The existing facilities are given as:

$$P_1 = (20, 10), P_2 = (8, 4), P_3 = (12, 6), and\ P_4 = (8, 24)$$

with the corresponding weights:

$$w_1 = 8, w_2 = 5, w_3 = 6, and\ w_4 = 1$$

Find a new facility location for which the total interaction cost at most is 25 percent greater than the minimum total interaction cost.

Solution

First, we determine the coordinates of the optimum location:

$$x^* = \frac{\sum_{i=1}^{m} w_i a_i}{\sum_{i=1}^{m} w_i} = 14.0$$

$$y^* = \frac{\sum_{i=1}^{m} w_i b_i}{\sum_{i=1}^{m} w_i} = 8.0$$

Next we compute the minimum interaction cost:

$$C_{min} = f\left(x^*, y^*\right) = \sum_{i=1}^{m} w_i \left[\left(x^* - a_i\right)^2 + \left(y^* - b_i\right)^2\right]$$

$$C_{min} = f(14,8) = \sum_{i=1}^{m} w_i \left[(14 - a_i)^2 + (8 - b_i)^2\right] = 920$$

The total cost being considered:

$$C_C = C_{min} + 0.25 C_{min} = 1,150$$

This is the cost if the new facility is located on the contour line. Within the circle, the cost will be less and will depend on the location.

Using the contour line radius equation:

$$r = \left[\frac{C_C}{W} + \left(x^*\right)^2 + \left(y^*\right)^2 - \sum_{i=1}^{m} \frac{w_i \left(a_i^2 + b_i^2\right)}{W} \right]^{1/2}$$

$$r = \left[\frac{1012}{20} + \left(14\right)^2 + \left(8\right)^2 - \sum_{i=1}^{m} \frac{w_i \left(a_i^2 + b_i^2\right)}{20} \right]^{1/2} = 3.4$$

Therefore, any point within or on a circle of 3.4 radius (Figure 8.3) is a location within the acceptable total cost.

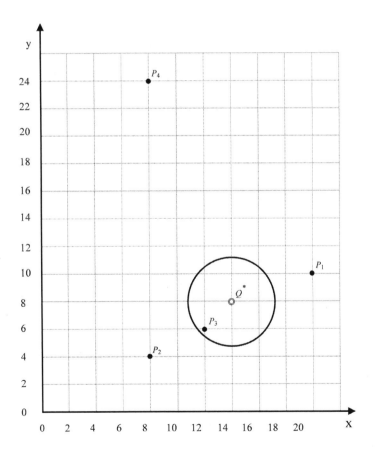

Figure 8.3. Solution for Example 1.

Observation

We notice that although cost is proportional to the square of the distance traveled, but since the interaction weights of the points P_1, P_2, and P_3 are significantly larger, the optimum new facility location is much closer to these points than to point P_4. With 25 percent greater interaction cost, a significant area becomes available, within which the new facility can be located. In fact, point P_3, which is within the contour line circle as an existing facility, if feasible, may be a convenient location for the new facility. In which case, the total annual interaction will be:

$$C_{new} = f(12,6) = \sum_{i=1}^{m} w_i \left[(12 - a_i)^2 + (6 - b_i)^2 \right] = 1080$$

This amount is about 17 percent more than the minimum cost.

8.4.2 Euclidean Distance

Following a similar formulation as with the Squared-Euclidean distance problems, the objective function for the Euclidean distance problem for a single new facility location can be obtained.

We defined the Euclidean distance between a new facility location and an existing facility as:

$$d_i(Q, P_i) = d_i = \left[(x - a_i)^2 + (y - b_i)^2 \right]^{\frac{1}{2}}$$

The objective function then is:

$$Minimize\ f(Q) = f(x, y) = \sum_{i=1}^{m} w_i d_i$$

or

$$Minimize\ f(Q) = f(x, y) = \sum_{i=1}^{m} w_i \left[(x - a_i)^2 + (y - b_i)^2 \right]^{\frac{1}{2}}$$

For these forms of minimization problems the standard mathematical method, as was shown in the previous section for the Squared-Euclidean problem, is to obtain the partial derivatives of the cost function with respect to the variables x and y, and on setting them equal to zero, determine the coordinates of the optimum new facility location. The partial derivatives are:

$$\frac{\partial f(x,y)}{\partial x} = \sum_{i=1}^{m} \frac{w_i(x - a_i)}{\left[(x - a_i)^2 + (y - b_i)^2\right]^{1/2}}$$

$$\frac{\partial f(x,y)}{\partial y} = \sum_{i=1}^{m} \frac{w_i(y - b_i)}{\left[(x - a_i)^2 + (y - b_i)^2\right]^{1/2}}$$

We notice that owing to the form of these equations, if the location of the new facility coincides with the location of an existing facility, one of the terms in each of the equations will become undefined as $\left(\frac{0}{0}\right)$. The only condition that allows continuing the procedure is that:

$$(x, y) \neq (a_i, b_i)\ i = 1 \dots m$$

But this cannot be readily verified when the new location is unknown. The way around this is to redefine the partial derivatives as proposed by Kuhn (1967).

If $(x, y) \neq (a_i, b_i)\ i = 1 \dots m$, it would be sufficient to solve the following set:

$$R(x, y) = \left(\frac{\partial f(x, y)}{\partial x},\ \frac{\partial f(x, y)}{\partial y}\right) = (0,0)$$

Otherwise, if $(x, y) = (a_k, b_k)$, $k = 1 \dots m$, the redefined modified gradient is:

$$R(x, y) = R(a_k, b_k) = \begin{cases} \left(\dfrac{u_k - w_k}{u_k} s_k,\ \dfrac{u_k - w_k}{u_k} t_k\right) & if\ u_k > w_k \\[2ex] (0,0) & if\ u_k \leq w_k \end{cases}$$

where w_k is the weight of each existing facility and:

$$s_k = \sum_{\substack{i=1 \\ \neq k}}^{m} \frac{w_i(a_k - a_i)}{\left[(a_k - a_i)^2 + (b_k - b_i)^2\right]^{1/2}}$$

$$t_k = \sum_{\substack{i=1 \\ \neq k}}^{m} \frac{w_i(b_k - b_i)}{\left[(a_k - a_i)^2 + (b_k - b_i)^2\right]^{1/2}}$$

$$u_k = \left(s_k^2 + t_k^2\right)^{\frac{1}{2}}$$

We may interpret s_k and t_k as follows. Since $P_k(a_k, b_k)$ is an existing point and is "suspected" to coincide with the solution, it will have no cost implications (a zero term in the cost function). The contribution of P_k to the gradients s_k and t_k is excluded with the $\neq k$ operator in the lower limits of the respective summation equation. This eliminates the unnecessary undefined term. Each term in the equation for s_k and t_k represents how "strongly" the solution is pulled to a particular location in the x and y coordinates, respectively. Then u_k is the resultant of s_k and t_k, that is, the collective pull of the weight of all existing facilities except P_k. Now, if for any point P_k, its weight, w_k, is greater (stronger) than u_k, then the modified gradient is $(0, 0)$. This means that point P_k pulls the optimum location to its location and coincides with it. Therefore, $Q(x^*, y^*) = Q(x, y) = P_k(a_k, b_k)$, and we stop here. Otherwise, if for all k, w_k is smaller (weaker) than u_k, then there is no possibility of an existing facility k to be the optimum location. The computation of quantities s_k, t_k, u_k is carried out for all existing points solely to verify the condition.

If $R(a_k, b_k) \neq (0, 0)$, we do not have a solution yet, but we now know that $(x, y) \neq (a_k, b_k)$.

We can now safely attempt to solve the partial derivative equations using the following procedure.

Setting $\dfrac{\partial f(x, y)}{\partial x}$ equal to zero we obtain:

$$x \sum_{i=1}^{m} \frac{w_i}{\left[\left(x - a_i\right)^2 + \left(y - b_i\right)^2\right]^{\frac{1}{2}}} = \sum_{i=1}^{m} \frac{w_i a_i}{\left[\left(x - a_i\right)^2 + \left(y - b_i\right)^2\right]^{\frac{1}{2}}}$$

Similarly, on setting $\dfrac{\partial f(x, y)}{\partial y}$ equal to zero we have:

$$y \sum_{i=1}^{m} \frac{w_i}{\left[\left(x - a_i\right)^2 + \left(y - b_i\right)^2\right]^{\frac{1}{2}}} = \sum_{i=1}^{m} \frac{w_i b_i}{\left[\left(x - a_i\right)^2 + \left(y - b_i\right)^2\right]^{\frac{1}{2}}}$$

For simplicity, letting:

$$g_i(x, y) = \frac{w_i}{\left[\left(x - a_i\right)^2 + \left(y - b_i\right)^2\right]^{\frac{1}{2}}}, i = 1, \dots m$$

Then x and y are determined as:

$$x = \frac{\sum_{i=1}^{m} a_i g_i(x, y)}{\sum_{i=1}^{m} g_i(x, y)}$$

and

$$y = \frac{\sum_{i=1}^{m} b_i g_i(x, y)}{\sum_{i=1}^{m} g_i(x, y)}$$

We cannot as yet consider these the final coordinates (x^*, y^*), because we see that the left-hand side of the above equations is a function of x and y, which are still unknown. We know that this form of equations can be solved through an iteration procedure. The iterative solution can be written as:

$$x^{(n)} = \frac{\sum_{i=1}^{m} a_i g_i\left(x^{(n-1)}, y^{(n-1)}\right)}{\sum_{i=1}^{m} g_i\left(x^{(n-1)}, y^{(n-1)}\right)}$$

and

$$y^{(n)} = \frac{\sum_{i=1}^{m} b_i g_i\left(x^{(n-1)}, y^{(n-1)}\right)}{\sum_{i=1}^{m} g_i\left(x^{(n-1)}, y^{(n-1)}\right)}$$

The superscripts indicate the iteration number. An initial guess or a starting seed value $(x^{(0)}, y^{(0)})$ is needed to run through the iteration process to estimate the solutions $(x^{(1)}, y^{(1)})$, $(x^{(2)}, y^{(2)})$, and so forth in succession until no appreciable change in the estimates occurs. Along the way care must be taken to ensure $g_i(x, y)$ is always defined; that is, the denominator in the equation that defines it does not become zero. If this occurs, to circumvent the problem, other starting seed values should be used.

The iteration termination conditions can be set as follows:

$$\left| x^{(n)} - x^{(n-1)} \right| \leq a$$

and

$$\left| y^{(n)} - y^{(n-1)} \right| \leq b$$

where a and b are acceptable tolerances with the unit used for the distance.

Typically, the center-of-gravity (Squared-Euclidean) solution is used as the starting point $(x^{(0)}, y^{(0)})$ for the iteration procedure, but any reasonable assumption will lead to convergence and the solution.

Example 8.2

A new facility is to be located among three existing facilities. The interaction between the new facility and the existing facilities is such that the cost is proportional to the Euclidean distance. The existing facilities are given as:

$$P_1 = (4,0), P_2 = (8,3), and\ P_3 = (2,6)$$

with the corresponding weights as:

$$w_1 = 2, w_2 = 10, and\ w_3 = 5$$

Determine the optimum location for the new facility.

Solution

We use Kuhn's modified gradient method to verify whether or not the new facility location coincides with the location of an existing facility.

The respective equations were:

$$s_k = \sum_{\substack{i=1 \\ \neq k}}^{m} \frac{w_i(a_k - a_i)}{\left[(a_k - a_i)^2 + (b_k - b_i)^2\right]^{1/2}}$$

$$t_k = \sum_{\substack{i=1 \\ \neq k}}^{m} \frac{w_i(b_k - b_i)}{\left[(a_k - a_i)^2 + (b_k - b_i)^2\right]^{1/2}}$$

$$u_k = \left(s_k^2 + t_k^2\right)^{1/2}$$

We now determine u_k, $k = 1, 2, 3$ and compare each, respectively with w_k, $k = 1, 2, 3$:

$$s_1 = \frac{10(4-8)}{\left[(4-8)^2 + (0-3)^2\right]^{1/2}} + \frac{5(4-2)}{\left[(4-2)^2 + (0-6)^2\right]^{1/2}} = -6.42$$

$$t_1 = \frac{10(0-3)}{\left[(4-8)^2 + (0-3)^2\right]^{1/2}} + \frac{5(0-6)}{\left[(4-2)^2 + (0-6)^2\right]^{1/2}} = -10.74$$

$$u_1 = \left[(-6.42)^2 + (-10.74)^2\right]^{1/2} = 12.5$$

Recall that $w_1 = 2$, and because $u_1 > w_1$, there is no possibility that the optimum new facility location coincides with the existing facility P_1. Therefore, we repeat the calculations for the next existing facility, P_2:

$$s_2 = \frac{2(8-4)}{\left[(8-4)^2 + (3-0)^2\right]^{1/2}} + \frac{5(8-2)}{\left[(8-2)^2 + (3-6)^2\right]^{1/2}} = 6.07$$

$$t_2 = \frac{2(3-0)}{\left[(8-4)^2 + (3-0)^2\right]^{1/2}} + \frac{5(3-6)}{\left[(8-2)^2 + (3-6)^2\right]^{1/2}} = -1.04$$

$$u_2 = \left[(6.07)^2 + (-1.04)^2\right]^{1/2} = 6.16$$

Knowing that $w_2 = 10$, and since $w_2 > u_2$, therefore, the location of the optimum new facility coincides with the location of the existing facility P_2. Thus optimum location for the new facility is:

$$(x^*, y^*) = (8, 3)$$

Because the new facility location has been found and confirmed, there is no need to continue the procedure to calculate u_3 for the third existing facility, P_3.

Graphical Solutions to Simple Cases

For a few simple Euclidean distance problems there exist geometric-graphical solutions that provide some intuitive understanding of the problems. These simple cases and their solutions are shown in Figure 8.4.

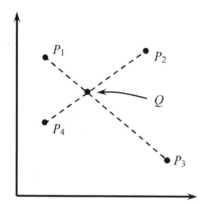

Four existing point facilities.
Equal weights.
Optimum new facility
location: Intersection point
of lines $\overline{P_1 P_3}$ and $\overline{P_2 P_4}$.

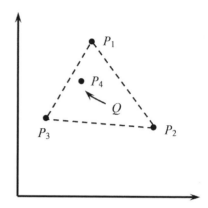

Four existing point facilities.
Equal weights.
Optimum new facility
location: Coincidental with
location of the existing
facility circumscribed by the
triangle formed by the other
three existing facilities.

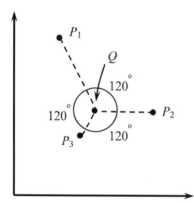

Three existing point facilities.
Equal weights.
Optimum new facility location:
The point within the triangle,
formed by the three points, that
makes 120° angle with each pair
of the points.

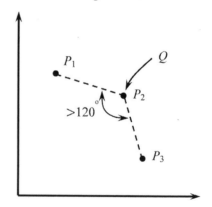

Three existing point facilities.
Equal weights.
Optimum new facility location:
Coincidental with facility P_2.

Figure 8.4. Graphical solutions to simple Euclidean distance problems.

Example 8.3

There are four existing facilities located as follows:

$$P_1 = (0, 0), P_2 = (2, 8), P_3 = (5, 1) \text{ and } P_4 = (6, 10)$$

A new facility is to be installed which will have equal interaction with the existing facilities. The cost is known to be proportional to Euclidean distance. It has been verified that the location of the new facility does not coincide with the location of any existing facility. Determine the location of the new facility.

Solution

Since it is known that the location of the new facility does not coincide with the location of an existing facility, we proceed to find the optimum location using the iteration method.

First, we determine the gravity (Squared-Euclidean) solution to use as the starting point:

$$x^{(0)} = x^* = \frac{\sum_{i=1}^{m} w_i a_i}{\sum_{i=1}^{m} w_i} = 3.25$$

$$y^{(0)} = y^* = \frac{\sum_{i=1}^{m} w_i b_i}{\sum_{i=1}^{m} w_i} = 4.75$$

Using the equations:

$$x^{(n)} = \frac{\sum_{i=1}^{m} a_i g_i \left(x^{(n-1)}, y^{(n-1)} \right)}{\sum_{i=1}^{m} g_i \left(x^{(n-1)}, y^{(n-1)} \right)}$$

and

$$y^{(n)} = \frac{\sum_{i=1}^{m} b_i g_i \left(x^{(n-1)}, y^{(n-1)} \right)}{\sum_{i=1}^{m} g_i \left(x^{(n-1)}, y^{(n-1)} \right)}$$

the iterations result in the following numerical values.

Iteration #1	$(x^{(1)}, y^{(1)}) = (3.208, 4.851)$
Iteration #2	$(x^{(2)}, y^{(2)}) = (3.195, 4.933)$
Iteration #3	$(x^{(3)}, y^{(3)}) = (3.189, 5.000)$

...

...

Iteration #12 $\qquad$ $(x^{(12)}, y^{(12)}) = (3.170, 5.237)$

Iteration #13 $\qquad$ $(x^{(13)}, y^{(13)}) = (3.170, 5.245)$

If the accuracies $\left| x^{(13)} - x^{(12)} \right| < 0.01$, and $\left| y^{(13)} - y^{(12)} \right| < 0.01$ are acceptable, then the iteration can be terminated, and we take:

$$\left(x^*, y^* \right) = \left(x^{(13)}, y^{(13)} \right) = (3.170, 5.245)$$

as the optimum location for the new facility. Otherwise, we continue the iteration until the desired accuracy is reached.

Iso-Cost Contour Lines

Using contour lines can provide many alternative near-optimal solutions when the optimum location is unavailable or inaccessible. We have the minimum cost due to the optimum location as:

$$C_{min} = f\left(Q^* \right) = f\left(x^*, y^* \right) = \sum_{i=1}^{m} w_i \left[\left(x^* - a_i \right)^2 + \left(y^* - b_i \right)^2 \right]^{1/2}$$

For any C_c, where $C_c > C_{min}$ the equation:

$$C_C = f(x, y) = \sum_{i=1}^{m} w_i \left[\left(x - a_i \right)^2 + \left(y - b_i \right)^2 \right]^{1/2}$$

yields the set of points (x, y), where the cost is fixed at C_c. However, except for a few simple cases, there is no straightforward procedure for constructing the contour lines for Euclidean distance location problems.

For $m = 1$ and any weight, the optimum location for the new facility coincides with location of the existing facility. We place the coordinate system at this common point, that is, at:

$$Q^* = P_1 = (0, 0)$$

Then the cost equation becomes:

$$C_C = f\left(x, y \right) = w_1 \left[\left(x \right)^2 + \left(y \right)^2 \right]^{1/2}$$

or

$$(x)^2 + (y)^2 = \left(\frac{C_C}{w_1}\right)^2$$

This is the definition of a circle, and the contour lines for various values of C_c are concentric circles around the optimum location, as shown in Figure 8.5(a).

For another simple case when $m = 2$ and equal weights, the contour lines for various values of C_c will be ellipses with focal points at the location of the two existing facilities a shown in Figure 8.5(b).

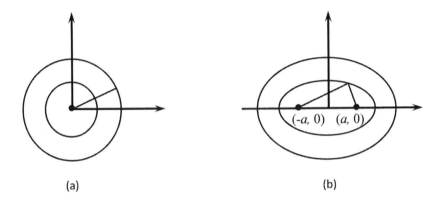

(a)	(b)

Figure 8.5. Contour lines for two simple Euclidean distance location problems.

To derive an easy-to-use mathematical relation for the contour lines, we place the x-axis along the line connecting the two points, locate the y-axis halfway between these points, and reorient the new axes to the final position shown in Figure 8.5(b).

Without loss of generality, we denote the transformed coordinates of the two existing facilities as:

$$P_1 = (a, 0) \ and \ P_2 = (-a, 0)$$

We can rewrite the iso-cost equation in terms of these points:

$$C_C = f(x, y) = w\left[(x-a)^2 + (y)^2\right]^{\frac{1}{2}} + w\left[(x+a)^2 + (y)^2\right]^{\frac{1}{2}}$$

$$\underbrace{\left[(x-a)^2 + (y)^2\right]^{\frac{1}{2}}}_{d_1} + \underbrace{\left[(x+a)^2 + (y)^2\right]^{\frac{1}{2}}}_{d_2} = \frac{C_C}{w}$$

The geometric relation $d_1 + d_2 = \dfrac{C_C}{w}$ (where $\dfrac{C_C}{w}$ is a constant) is the definition of an ellipse.

For $m \geq 3$ and any values of w_i there is no regular-shaped or predictable pattern for the contour lines. However, for any C_c ($C_c > C_{min}$) the contour lines can be numerically generated and plotted using most mathematical software, such as MATLAB®. The iso-cost contour lines will be irregular, smoothly varying, closed loop curves. A larger loop will signify a higher value of C_c.

8.4.3 Rectilinear Distance—Point Locations

Earlier in this chapter, we briefly introduced the concept of rectilinear distance. The rectilinear distance between two points $Q(x, y)$ and $P_i(a_i, b_i)$ is defined as:

$$d_i(Q, P_i) = d_i = |x - a_i| + |y - b_i|$$

If the interaction, transportation or movement between P_i, $i = 1, ..., m$ existing facilities and a new facility being considered occur through the rectilinear distance, the objective function for the optimum location of the new facility is:

$$Minimize \; f(Q) = f(x, y) = \sum_{i=1}^{m} w_i d_i$$

or

$$Minimize \; f(Q) = f(x, y) = \sum_{i=1}^{m} w_i |x - a_i| + \sum_{i=1}^{m} w_i |y - b_i|$$

Because the two terms in this equation are positive, each can be minimized separately. The mathematical features of each summation term, for example, the first term, leads to a solution where the optimum x coordinate of the new facility:

i.　Will be the same as the x coordinate of an existing facility.

ii.　This x coordinate will be such that less than one half of the total weight will be to its left, and less than one half will be to its right. This can also be said in a different way, that the x coordinate is such that the difference between the partial total weight to the left of x and the partial total weight to the right of x (as a positive quantity) is the smallest possible.

This solution is also referred to as the "median condition." The same solution approach holds for the y coordinate of the optimum new facility. This method can be readily illustrated using a numerical example.

Example 8.4

There are five existing service centers in a city, where a distribution warehouse is to be optimally located among them as a supply point. The movement between the warehouse and the service centers will be along rectilinear paths. The coordinate of the outlets and their respective interaction weights are given as:

$$P_1 = (1, 1), P_2 = (3, 5), P_3 = (4, 7), P_4 = (6, 8) \text{ and } P_5 = (8, 2)$$

and

$$w_1 = 4, w_2 = 3, w_3 = 4, w_4 = 8, \text{ and } w_5 = 6$$

Solution

Recall that optimum x-coordinate and optimum y-coordinate can be determined separately. We construct a table such as Table 8.1. The key point to remember is that the coordinate values in the second column must be either in ascending or descending order. The total interaction weight is 25, and its half (referred to as the half-sum) is 12.5.

Table 8.1. Solution procedure for the x-coordinate

Facility i	Coordinate (Ascending Order)	Weight W_i	Partial Sum $\sum_{j=1}^{i} w_j$
1	1	4	4
2	3	3	7<12.5
3	4	4	11<12.5
4	6	8	* 19>12.5
5	8	6	25>12.5

Referring to the last column, we locate the row where the inequality changes direction. This row corresponds to the x-coordinate of the existing facility P_4. Therefore:

$$x^* = a_4 = 6$$

Table 8.2 shows the same process for the y-coordinate.

Table 8.2. **Solution procedure for the y-coordinate**

Facility i	Coordinate (Ascending Order)	Weight W_i	Partial Sum $\sum_{j=1}^{i} w_j$
1	1	4	4
5	2	6	10<12.5
2	5	3	* 13>12.5
3	7	4	17>12.5
4	8	8	25>12.5

We note that the inequality sign changes direction at the y-coordinate of the existing facility P_2. Therefore:

$$y^* = b_2 = 5$$

Although the x^*- and y^*-coordinates in this case come from the coordinates of two different existing facilities, as outlined earlier, it is entirely possible that both coordinates are from a single existing facility. It is a matter of how the numerical quantities work out.

We now introduce an alternative method of finding the optimum new facility location. This will not only give a visual understanding of the solution, it will also provide a means to construct iso-cost contour lines for the rectilinear facility location problems. In this method, the weights are likened to forces acting on a beam spanning the maximum distance between the existing facilities in their x and y coordinates. Using the previous example, the concept of this solution approach is shown in Figure 8.6. We assume a freestanding beam and a single moving support, as the x coordinate of the new facility, along the x-axis from the left of the left-most existing facility to the right of the right-most existing facility. If the new facility is to be located at $x < 1$, there will be a pulling force of 0 to its left, and a cumulative pulling force 25 to its right. With the sign convention shown in Figure 8.6, the resultant force will be −25. That is, the resultant force, tending to rotate the beam clockwise with respect to the location of support, is assumed to be negative. (Similarly, we can interpret the resultant force pulling the support toward the right.) Now we begin to move the support to the right in a step-wise manner and examine the balance of forces with respect to its location. When $1 < x < 3$, there will be a cumulative force of 4 to its left and a cumulative force of 21 to its right. The resultant force is $4 - 21 = -17$. We continue moving the support to the right to the end point. The steplike plot of the resultant force at one point crosses the x-axis. This point is the solution $x^* = 6$.

The same result can also be deduced from the vertical axis on the right, where the step-wise plot, corresponding to the cumulative force to the right of x crosses the x-axis at $x = x^* = 6$.

The similar procedure for the y coordinate of the new facility is shown in Figure 8.7, where $y^* = 5$.

Iso-Cost Contour Lines

Numerical findings from Figures 8.6 and 8.7 can be combined to construct the diagram in Figure 8.8. We denote the resultant forces for the x and y-axes by M_i and N_j, respectively. These values are also shown in Figure 8.8 over the segments

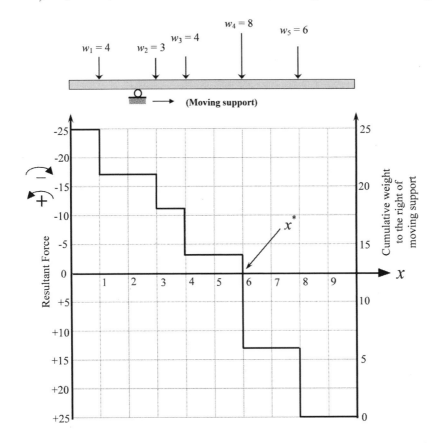

Figure 8.6. Force analogy for x-axis.

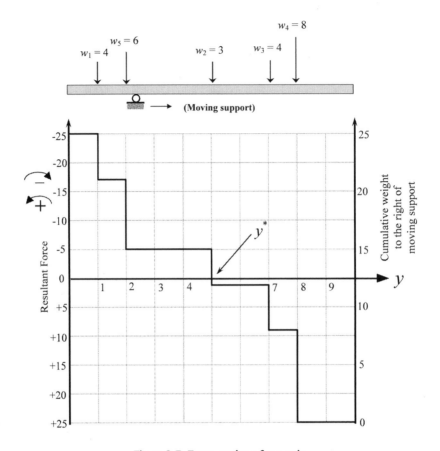

Figure 8.7. **Force analogy for y-axis.**

that they apply. The resultant of these force components pulls the new facility location toward the optimum location. The force components and resultant are different for each rectangular region bounded by the x and y coordinates of the existing facilities and always point to the optimum location. The slope of the resultant force for each region is:

$$R_{ij} = \frac{N_j}{M_i}$$

In fact, R_{ij} represent the steepest descend for the cost function $f(Q)$ and are shown within the respective rectangular regions in Figure 8.8. In contrast to the direction of the steepest descent, if we consider a direction perpendicular to it, we will stay on a constant value of the cost function, or zero slope. Drawing these zero-slope direction lines from any arbitrary point and continuing from the boundary of each region to the next, in an end-to-end fashion, results in a closed loop path, or the so-called iso-cost contour line. Denoting the direction of a zero slope as S_{ij}, for the two perpendicular lines we have:

$$R_{ij} \cdot S_{ij} = -1$$

or

$$S_{ij} = -\frac{1}{R_{ij}} = -\frac{M_i}{N_j}$$

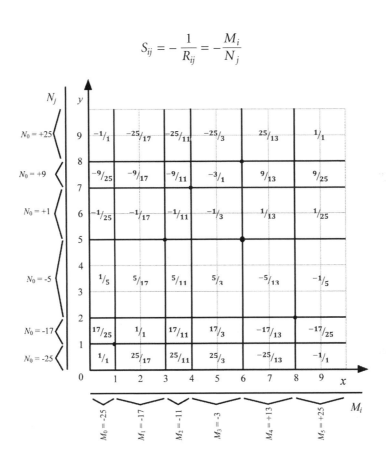

Figure 8.8. Steepest descent directions

All the closed paths shown in Figure 8.9 are the iso-cost contour lines for various values of the cost function. The larger the path, the higher the cost.

8.4.4 Rectilinear Distance—Area Locations

The assumption that existing facilities are located at a point can be a good approximation when the dimensions of the area occupied by the facility are

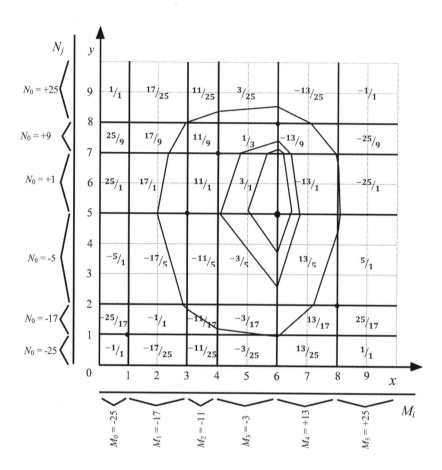

Figure 8.9. Contour lines

much smaller than the distance between the centers of the facilities. If the size of the areas to be served by the new facility, such as a warehouse that supplies large geographical regions, then the assumption of point location for the existing facilities will be inappropriate to make. It is possible to extend the solution of point-location problems to area-location problems when the interaction weights are uniformly distributed over the respective areas. To demonstrate this, we consider an example involving both existing point and area locations.

Example 8.5

The locations of three point and four area facilities are shown in Figure 8.10, and the weights for the facilities are given in Table 8.3. A new facility is to be located relative to the existing facilities with rectilinear distance travel. Determine the location of the new facility.

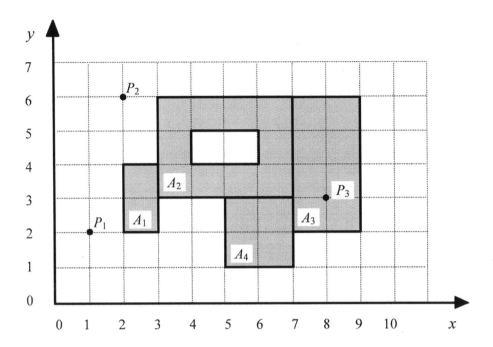

Figure 8.10. Existing facility locations.

Table 8.3. Weights for the existing facilities

Point	Weight
1	3
2	4
3	1

Area	Weight
1	4
2	5
3	4
4	4

Solution

Area weights are denoted by v_j. Similar to the case of the point location problem, we begin by plotting the cumulative weight to the right of x for the range $0 \leq x \leq 9$, as shown in Figure 8.11. The plot of cumulative weight to the right of x crosses the x-axis at $x = x^* = 4$. Similarly, for the y-axis this occurs at $y = y^* = 3.3$, as shown in Figure 8.12.

The numerical value of $y^* = 3.3$ is determined from interpolation between end points of the line segment between $y = 3$ and $y = 4$. The key in plotting the diagrams of Figures 8.11 and 8.12 is in considering uniformly distributed weight

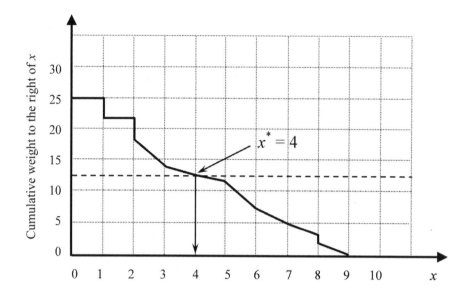

Figure 8.11. Weight distribution for *x*-axis.

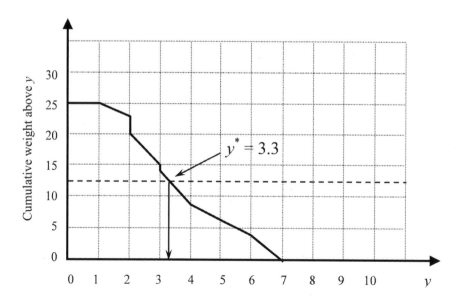

Figure 8.12. Weight distribution for *y*-axis.

over the respective areas. For instance, in Figure 8.12, traversing from $y = 2$ to $y = 3$, we cover the following distributed weights:

$$\frac{1}{2} v_1 = 2 \ for \ A_1$$

$$\frac{2}{4} v_4 = 2 \ for \ A_4$$

$$\frac{2}{8} v_3 = 1 \ for \ A_3$$

This represents a total of 5, which correspond to cumulative drop from 20 at $y = 2$ to 15 at $y = 3$.

An irregularly shaped area location can be approximated by dividing it into many small squares, counting the total number of squares, and dividing the weight for the area by the total number of squares and obtaining a weight per square. Then the problem can be solved in a similar way as in the example above.

8.5 MINI-MAX LOCATION PROBLEMS

In this class of problems, the objective function is formulated as:

$$Minimize \ f(Q) = f(x, y) = Max \ d_i(Q, P_i), \ i = 1, m$$

where Q denotes the location of the new facility, and d_i is the distance between Q and an existing facility P_i.

In the case of the rectilinear distance this becomes:

$$Minimize \ f(Q) = f(x, y) = Max \ \left[|x - a_i| + |y - b_i| \right], \ i = 1, m$$

With no proof presented here, the optimal solution is obtained by computing a number of parameters as follows:

$$C_1 = minimum \ (a_i + b_i)$$

$$C_2 = maximum \ (a_i + b_i)$$

$$C_3 = minimum \ (-a_i + b_i)$$

$$C_4 = maximum \ (-a_i + b_i)$$

$$C_5 = maximum \ (C_2 - C_1, C_4 - C_3)$$

Then, the optimum solution encompasses all the points on the line segment connecting the points:

$$\left(x^*, y^*\right) = \frac{1}{2}\left(C_1 - C_3, C_1 + C_3 + C_5\right)$$

and

$$\left(x^{**}, y^{**}\right) = \frac{1}{2}\left(C_2 - C_4, C_2 + C_4 - C_5\right)$$

The minimized maximum distance is:

$$d_{min} = \frac{C_5}{2}$$

The line segment will be either a 45° (*slope* = 1) or a 135° (*slope* = −1) line in the $x - y$ coordinate system. Access to any point on the solution line from the existing facilities must be in rectilinear form.

Iso-Distance Contour Lines

The rectilinear iso-distance contour lines or boundary around a point, or for the general case of a 45° or a 135° line, are drawn as shown in Figure 8.13. Any point on any rectangle has the same rectilinear distance from the inclined line that is

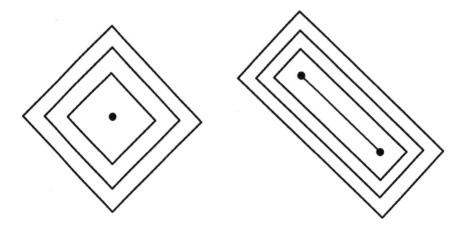

Figure 8.13. Mini-max iso-distance contour lines.

the optimum location. If the optimum location happens to be a point, then the rectangles become squares.

Example 8.6

The coordinates of eight machining centers expressed in units of blocks traveled are given in Table 8.4. To provide tools and supplies for the machining centers, a central storage system is being considered. The machines are laid out such that the access is through rectilinear movement. It is desired that location of new storage be the most accessible to all machining centers.

a.   Determine the optimal location for the storage system.

b.   If it turns out that the new optimal location is inaccessible, it is acceptable to travel two additional blocks. Draw the iso-distance contour line.

Table 8.4. Coordinates of the machining centers

i	a_i	b_i		i	a_i	b_i
1	1	2		5	6	3
2	2	8		6	7	7
3	4	7		7	9	4
4	5	1		8	9	8

Solution

(a).   Using the relations for the mini-max problem, the relevant numerical calculations are shown in Table 8.5.

Table 8.5. Calculation data

i	a_i	b_i	$a_i + b_i$	$-a_i + b_i$
1	1	2	3	1
2	2	7	9	5
3	3	3	6	0
4	4	7	11	-3
5	5	2	7	-3
6	7	7	14	0
7	8	5	13	-3
8	9	8	17	0

From Table 8.5:

$$C_1 = 3, C_2 = 17, C_3 = -3, C_4 = 5, \text{ and } C_5 = 14$$

The optimum solution is all the points on the line segment connecting the points:

$$(x^*, y^*) = (3, 7)$$

and

$$(x^{**}, y^{**}) = (6, 4)$$

The solution is shown in Figure 8.14. The minimized maximum distance is:

$$d_{min} = 7 \ [blocks]$$

(b). The contour line for the additional two blocks is also shown in Figure 8.14. The maximum distance from any machining center will be no more than $7 + 2 = 9 \ [blocks]$. The storage system must be located at an available point within or on the boundary of the rectangle.

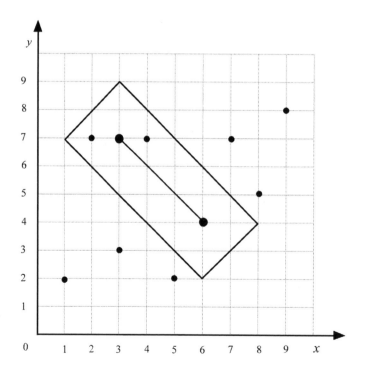

Figure 8.14. **Solution and iso-distance contour line.**

EXERCISES

8-1 What is the number one objective in location problems?

8-2 What are other objectives of interest is location problems?

8-3 Why can the x and y coordinates of the rectilinear location problems be determined independent of each other?

8-4 There are four existing facilities with the coordinates $P_1 = (1, 2)$, $P_2 = (3, 7)$ $P_3 = (4, 6)$, *and* $P_4 = (6, 4)$. A new facility is to be located relative to these facilities, where the item movement will be in straight lines. The relative weights of the annual item movement are $w_1 = 3$, $w_2 = 4$, $w_3 = 6$, *and* $w_4 = 8$.

   a) Determine the optimum location for the new facility if the cost is proportional to the square of the distance traveled. (Squared-Euclidean distance problem.)

   b) Determine the optimum location for the new facility if the cost is assumed to be proportional to the straight line traveled. (Euclidean distance problem.)

8-5 A new facility is to be located among six existing facilities and supply them with their raw material needs through rectilinear distance transportation. The relative locations of the existing facilities are $P_1 = (0, 0)$, $P_2 = (2, 3)$, $P_3 = (2, 5)$, $P_4 = (3, 7)$, $P_5 = (5, 5)$ *and* $P_6 = (6, 4)$. The interaction weights between the new and each of the existing facilities are $w_1 = 3$, $w_2 = 3$, $w_3 = 5$, $w_4 = 4$, $w_5 = 2$, *and* $w_6 = 6$.

   a) Determine the location of the new facility that minimizes the total weighted interaction effort.

   b) If the new facility cannot be located closer than 2 rectilinear distance to any existing facility, using only a geometric graph, make a prediction of the optimum location. (This is only a guess, and the result may not be accurate.)

   c) In follow-up to (b), determine optimum location with the aid of iso-cost contour lines. Compare the result with the prediction made in (b).

8-6 A new facility is to be located to serve four points and three area locations. The existing facilities are shown in Figure 8.15, and their weights are given in Table 8.6. The weights are uniformly distributed over the areas, and rectilinear distance is used.

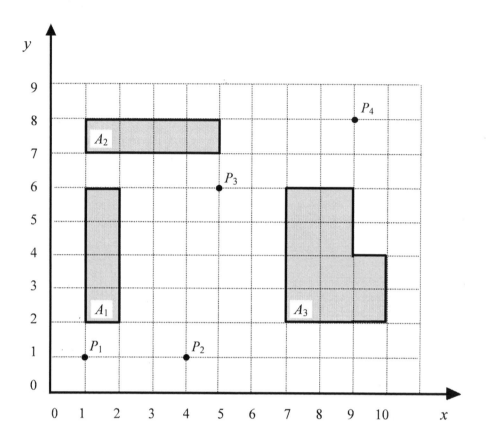

Figure 8.15. Existing facility locations.

Table 8.6. Weights for the existing facilities

Point	Weight
1	2
2	3
3	4
4	6

Area	Weight
1	6
2	4
3	10

Determine the mini-sum optimum location for the new facility.

Chapter 9 SYSTEM RELIABILITY

9.1 INTRODUCTION

The general concept of reliability is extremely wide, and reliability is a concern in almost every system that one can think of, from a simple device to most complex engineering systems. Reliability directly affects safety and availability. Lack of safety and unavailability can lead to a host of issues ranging from bodily harm to loss of productivity, or from frustration to damage to reputation for quality.

How many times have you wanted to use something or have been using something when a breakdown has dampened your spirit? Think about automobile problems, computer malfunctions, and the like in everyday life. Extend this to the operation of a transit system, hospital and emergency equipment, machinery, and the production facility, and the consequences of unreliability can be costly. Unreliability is generally associated with and is the result of a failure of a component or a module in the system. In theory, well-designed, well-engineered, thoroughly tested, and properly maintained equipment or systems should not fail in operation. But experience has shown that no amount of care can completely eliminate the chances of such failures from occurring. We can, however, by controlling the environmental effects, through design, and proper instructions for use, significantly influence and improve the reliability of a system.

9.2 DEFINITION OF RELIABILITY

Reliability is the capability of a device or a system to perform its intended function without breakdown.

We can formalize the reliability of equipment and systems based on the reliability of their individual modules and components and analyze them. The component level of the study of reliability is primarily the concern of design engineers, such as improving the reliability or the probability of a successful operation of a device or a sensor. But the reliability of larger systems, as integrated units to perform as a whole, can be an issue for industrial engineers. With this in mind, after reviewing the preliminaries of relevant functions and factors, we limit ourselves to reliability analysis of systems for which we know the probability of success for each of its constituent components. Such systems could include manufacturing lines, material handling and transportation systems, and, in fact, any interconnected system. For any meaningful analysis to be carried out, we need to know at least three elements:

1. *The reliability or probability of success of the operation of the individual components or modules.* These primarily come from recorded and historical data, experience, scientific or previous reliability analysis, claims by manufacturers, and, if none of the preceding is available, from intuition and guess. With the latter two, anything concluded will be an approximation and must be treated cautiously.

2. *How the individual components and modules are interconnected to process an input and direct it to the output.* The function of the elements of the system and the nature of inputs and outputs can relate to entities such as electricity, steam, parts in a manufacturing plant, people, equipment, or sensors in a control loop.

3. *The methods and means of determining the reliability of the entire system, and how we can influence it.* A major purpose of a reliability analysis often is to identify weak links and modules in terms of their individual reliability and their role in the total reliability of the system.

9.3 FAILURE OVER THE OPERATING LIFE

The failure rate or, more precisely, the instantaneous failure rate of many components and systems over their operating life can be divided into three phases or periods, as shown in Figure 9.1.

- Period *A*. This is called the early failure, run-in, burn-in or debugging phase. As the time passes for a new system, the failure rate decreases both naturally and as a result of fixes and fine-tunings performed.

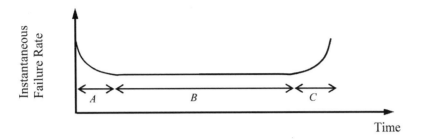

Figure 9.1. Failure rate over lifetime.

- Period *B*. This constitutes the extended operating life, and it is appropriately called the "useful life." In this phase there are occasional failures and breakdowns that are mostly random in nature—or by chance, although in some circumstances a failure can be related to the construct of the system and how it is used. Based on economic considerations, the nature and function of the components of the system, the operating instruction for a system, or the philosophy of an organization, the system may require some form of service and maintenance to be in place.
- Period *C*. This is the phase where increasingly frequent failures affect the operation of the system. Known as aging and the wearing-out period, repeated failures of different components of the system hamper the successful operation of the system. Not only can operating a system in this range be dangerous; it can also be very costly to maintain. Relying on an economic analysis, a major overhaul, and replacement of certain critical and aging components can move the system back into region *B* for a period of time.

9.4 THE RELIABILITY FUNCTION

In Section 9.2, we defined *reliability* as the probability that the system will perform the intended tasks satisfactorily and without failure. This should ideally be stated over a given period of time and under specific operating conditions. However, typically the operating period is envisaged to be the operating cycle (or expected lifetime), and the operating conditions are assumed to be reasonable and similar to the conditions under which the reliability measure had been determined. For this reason, it is common to express the reliability of most systems, components, or products simply as a numerical value.

The reliability function is defined as:

$$R(t) = 1 - F(t)$$

Where $F(t)$ is the probability that the system might fail by time t. $F(t)$ is essentially the unreliability or the failure distribution function. Failures occur randomly in time; if we show the probability density function as $f(t)$, we can write:

$$Probability\ of\ failure\ before\ time\ t = F(t) = \int_{-\infty}^{t} f(t)$$

Substituting this in the above equation:

$$R(t) = 1 - \int_{-\infty}^{t} f(t) = \int_{t}^{\infty} f(t)$$

Function $f(t)$ can take different forms for various systems, components and products. One common form that applies to many systems, but not all, is the "negative exponential," which is often simply, but not desirably, referred to as "exponential." For the sake of the derivations, we only consider the exponential density function, which is defined as:

$$f(t) = \lambda e^{-\lambda t}$$

where λ is the failure rate, and t is the operating time of interest. Then:

$$R(t) = \int_{t}^{\infty} \lambda e^{-\lambda t} = e^{-\lambda t}$$

The mean life is defined as:

$$\theta = \frac{1}{\lambda}$$

For the exponential function, this is more appropriately referred to at the Mean Time between Failures (MTBF).

To be able to determine the reliability (probability of success) over the time period t, we need the failure rate λ.

9.4.1 Failure Rate

The failure rate is the frequency with which a system or a component fails, and it is expressed as:

$$\lambda = \frac{number\ of\ failures}{total\ operating\ time}$$

The failure rate may be expressed in any unit of time depending on the system, but failure per hour, although numerically may appear small, is common. The reason is that $\frac{1}{\lambda}$, the MTBF, is then expressed in tangible units of "hour." The failure rate can be obtained in a number of ways: from past data of similar components, recorded data, specific tests conducted on a set of components, or, as a last resort, from intuition and experience—in which case rather than estimating the failure rate, it may be more useful to directly express the reliability R.

We now review in detail an example of determining the failure rate. Suppose that a total of ten components were concurrently put to the test under typical operating conditions. The failure data are given in the first column of Table 9.1. In this experiment after the failure of the fifth component at 480 hours, the relevant computation is shown in the rest of the table.

Table 9.1. Component testing failure data

Component Failure Times [hr]	Number of Components Operating	Time to Next Failure [hr]	Partial Operating Time [hr]
0	10	60 – 0 = 60	10 x 60 = 600
60	9	105 – 60 = 45	9 x 45 = 405
105	8	125 – 105 = 20	8 x 20 = 160
125	7	295 – 125 = 170	7 x 170 = 1,190
295	6	490 – 295 = 195	6 x 195 = 1,170
480	5	--	--
		Total	3,525

The five failures occurred over a total operating time of 3,525 hours. Therefore, the failure rate is:

$$\lambda = \frac{5}{3525} = 0.001418 \left[\frac{1}{hr}\right]$$

The MTBF is determined as:

$$\theta = \frac{1}{\lambda} = 705 [hr] \approx 1 [month]$$

9.5 RELIABILITY OF MULTIUNIT SYSTEMS

We designate the reliability or the probability of successful operation of a component by p, and its unreliability or probability of failure by its complement

$(1 - p)$ or q. We show the interconnection of components, modules (a group of components), and subsystems in a graphical form, and assume a certain direction of flow from the input, often shown on the left, to the output shown on the right. In the following sections, we review various forms of component and module interconnection and the way to obtain the reliability of a system from the reliability of its elements.

9.5.1 Series Configuration

A series configuration is shown in Figure 9.2.

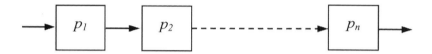

Figure 9.2. **Units in series connection.**

If the probability of successful operation of ith component is denoted by p_i, taken commonly as its reliability rating, for a series configuration the system reliability R_{system}, or simply R_s, is defined as:

$$R_S = p_1 \cdot p_2 \cdot p_3 \cdots p_n$$

$$R_S = \prod_{i=1}^{n} p_i$$

If $p_i = p$, *same for all i*, then $R_s = p^n$.

This is the simplest form of configurations. In fact, a major part of many systems is integrated or assembled in this form. Manufacturing lines where a succession of processes must be performed represent a serial configuration. Clearly, if any of the components in such a system becomes dysfunctional, its reliability becomes zero. Consequently, the entire system reliability also becomes zero, meaning that the system will fail to accomplish its function. This is a major concern in the serially connected system. In practice, because the probability of a successful operation of any nontrivial component will be less than 100 percent, the addition of any component in series reduces the system reliability.

Example 9.1

A system is made of two modules in series. Each module has a reliability of 0.98. What is the reliability of the system? What would be the system reliability if a third module of the same type is added?

Solution

The modules are identical, so for two modules:

$$R_S = p^n = 0.98^2 = 0.960$$

For three modules:

$$R_S = p^n = 0.98^3 = 0.941$$

We note that even with modules of relatively high reliability, by increasing the number of modules in series, the system reliability diminishes quickly.

9.5.2 Parallel Configuration

The construct of any device or system may require both serial and parallel interconnections. The parallel configuration, however, is often a "choice" and not a requirement by design or intended function. Components or modules in parallel can be intentionally added to improve the reliability of the system. This action is commonly referred to as adding "redundancy" to the system. A parallel configuration is shown in Figure 9.3.

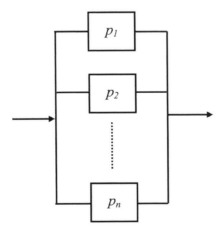

Figure 9.3. **Units in parallel connection.**

If the probability of successful operation of the ith component is represented by p_i, and the system can operate as long as at least one component is functional, then for a parallel configuration the system reliability R_s is given as:

$$R_S = 1 - probability\ that\ the\ system\ is\ in\ failed\ state$$

This is commonly referred to as the "unreliability law of parallel system." Mathematically:

$$R_S = 1 - (1 - p_1)(1 - p_{2)}) \dots (1 - p_n)$$

$$R_S = 1 - \prod_{i=1}^{n}(1 - p_i)$$

If $q_i = (1 - p_i)$:

$$R_S = 1 - \prod_{i=1}^{n} q_i$$

If $q_i = q$ *same for all* i, then $R_s = 1 - q^n$.

Essentially, in this configuration, if one component fails, other components continue or take over the intended operation of the system.

Example 9.2

A system is composed of a single module with a poor reliability value of only 0.75. To improve the system's reliability through redundancy, an identical module is added in parallel. What will the system's reliability be? What if another level of redundancy is added?

Solution

The redundancy is achieved through identical modules, so for two modules:

$$q = 1 - 0.75 = 0.25$$

$$R_S = 1 - 0.25^2 = 0.937$$

For three modules:

$$R_S = 1 - 0.25^3 = 0.984$$

We see that adding redundancy significantly improves the system reliability, and through more levels of redundancy, the system reliability will approach 100 percent.

9.6 COMBINED SERIES-PARALLEL CONFIGURATIONS

The general form of combined series-parallel configuration is shown in Figure 9.4. We examine two combined series-parallel configurations. Because such configurations also result from adding "redundant" modules to increase the overall system reliability, we refer to both as specific forms of redundancy.

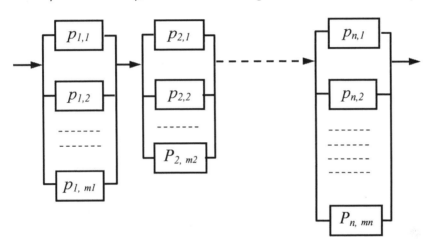

Figure 9.4. Combined series-parallel configuration.

9.6.1 System-Level Redundancy

To increase the reliability, there are situations where one or more complete systems are put in parallel with the original system. In principle, it is not necessary that they all be identical. The purpose of this type of redundancy is that when one system fails, others continue or take over so that the intended operation is accomplished. An emergency power generator is an example of system-level redundancy with nonidentical systems. Another example of complete duplication at the system level or adding redundancy is the assignment of a copilot to a pilot in flying an aircraft. The general form of redundancy at the system-level is shown in Figure 9.5.

For each series subsystem j:

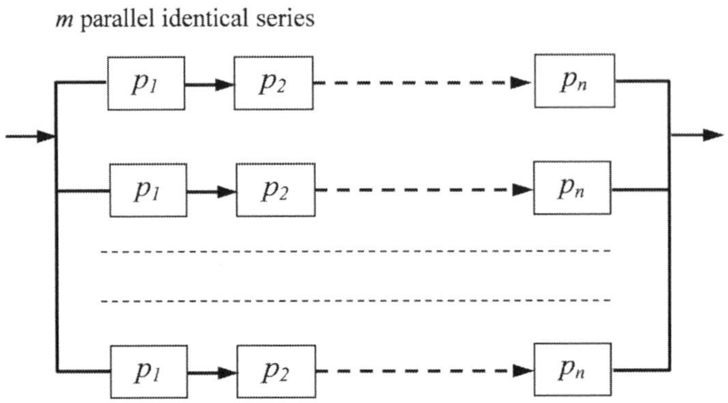

m parallel identical series

Figure 9.5. Redundancy at the system level.

$$R_j = \prod_{i=1}^{n} p_i \quad and \quad Q_j = 1 - R_j$$

For *m* parallel subsystems:

$$R_S = 1 - \prod_{j=1}^{m} Q_j$$

or

$$R_S = 1 - \prod_{j=1}^{m} (1 - \prod_{i=1}^{n} p_i)$$

9.6.2 Component-Level Redundancy

An alternative to system-level redundancy is to use the same number of each of the system's components and rearrange them in parallel. For example, the system of Figure 9.5 can be redesigned as in Figure 9.6.

At each stage, for *m* parallel modules we have:

$$R_i = 1 - \prod_{j=1}^{m} q_j$$

m identical modules/components in each stage

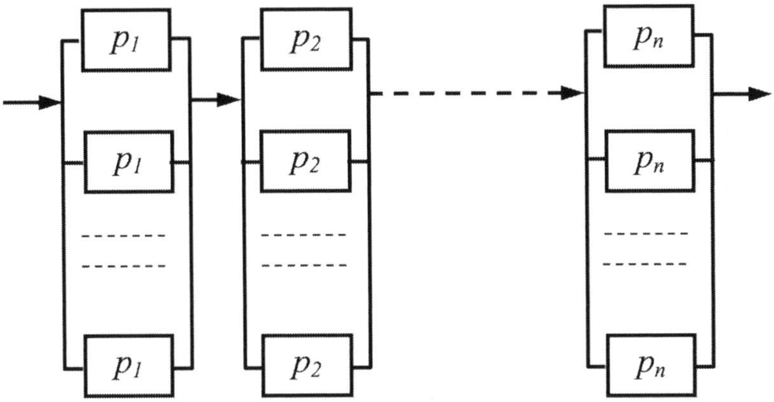

Figure 9.6. Redundancy at the component level.

For the *n* stages in series we obtain:

$$R_S = \prod_{i=1}^{n} R_i = \prod_{i=1}^{n}\left(1 - \prod_{j=1}^{m} q_j\right)$$

Example 9.3

Several families of components, F_i, are to be produced on four types of machines, *L*, *M*, *D*, and *G*. Four machines of each type are available, and two systems of physical arrangements have been proposed as reviewed in Chapter 3: functional layout, designated as *System I*, and group technology layout, designated as *System II*, as shown in Figures 9.7 and 9.8.

In *System I*, although each component family may use any of the machines within each group, during the normal operation of the system it has been found both practical and convenient for component family F_1 to use machines L_1, M_1, D_1, and G_1 on a permanent basis. Similarly, in *System II*, component family F_1 has been assigned to *Group* 1. During certain periods of time the component family received a high demand and because of this, it is given top priority. Under such circumstances other machines in the case of *System I* and other groups in the case of *System II* are immediately made available to the component family F_1 on an active redundancy basis:

(a)   Determine the reliability of each system as far as the production of the component family F_1 is concerned using the following reliability data:

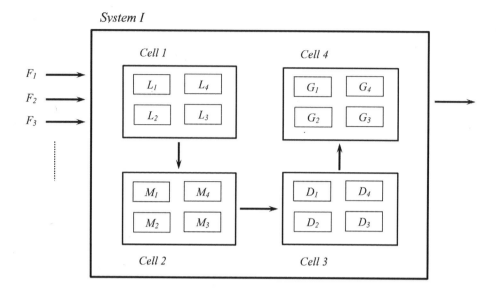

Figure 9.7. System I.

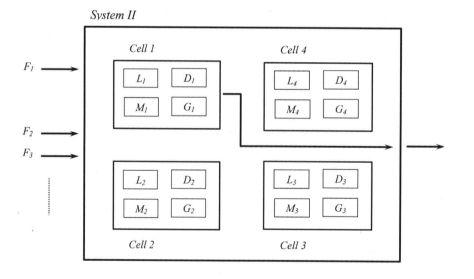

Figure 9.8. System II.

$$p_{L_i} = p_{D_i} = 0.8 \text{ and } p_{M_i} = p_{G_i} = 0.7, i = 1 \text{ to } 4$$

(b)   If the reliability measures are taken to mean availability for production, determine the expected production rate of the component family F_1 under the high-demand condition for each of the systems having the following production time data:

$$T_{L_i} = 0.5, T_{M_i} = 0.4, T_{D_i} = 0.1, \text{ and } T_{G_i} = 0.5 \text{ units of time per batch, } i = 1 \text{ to } 4$$

Solution

Part (a): We reproduce the two systems in block diagrams as shown in Figures 9.9 and 9.10.

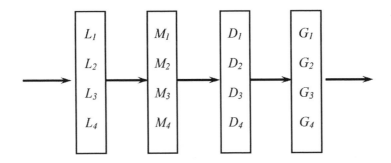

Figure 9.9. Block diagram of *System I.*

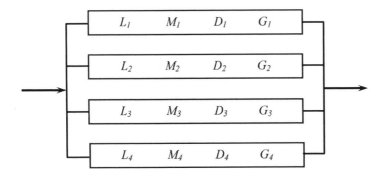

Figure 9.10. Block diagram of *System II.*

Stage reliabilities, $j = 1, 4$, for *System I*:

$$R_1 = R_3 = 1 - \prod_{i=1}^{4} q_i = 1 - \prod_{i=1}^{4}(1 - 0.8) = 0.9984$$

$$R_1 = R_4 = 1 - \prod_{i=1}^{4} q_i = 1 - \prod_{i=1}^{4}(1 - 0.7) = 0.9919$$

$$R_{System I} = \prod_{j=1}^{4} R_j = (0.9984)(0.9984)(0.9919)(0.9919)$$

$$R_{System\ I} = 0.9807$$

Line reliabilities, $j = 1, 4$ for *System II*:
Since all four lines are identical:

$$R_1 = R_2 = R_3 = R_4 = \prod_{i=1}^{4} p_1 = (0.8)(0.7)(0.8)(0.7) = 0.3136$$

$$Q_j = 1 - R_j = 1 - 0.3136 = 0.6864, \ j = 1 \ to \ 4$$

$$R_{System\ II} = 1 - (Q_j)^4 = 1 - (0.6864)^4$$

$$R_{System\ II} = 0.7780$$

The *System I* is notably more reliable than *System II*.

Part (b): In each of the systems one of the machines has the longest processing time corresponding to the lowest production rate. From the data given, we compute the production rate, r_k for each type of the machines:

$$r_L = \frac{1}{T_L} = \frac{1}{0.5} = 2 \left[\frac{batches}{unit\ time} \right]$$

$$r_M = \frac{1}{T_M} = \frac{1}{0.4} = 2.5 \left[\frac{batches}{unit\ time} \right]$$

$$r_D = \frac{1}{T_D} = \frac{1}{0.1} = 10 \left[\frac{batches}{unit\ time} \right]$$

$$r_G = \frac{1}{T_G} = \frac{1}{0.3} = 3.33 \left[\frac{batches}{unit\ time} \right]$$

The smallest production rate is $r_L = 2.0 \left[\frac{batches}{unit\ time} \right]$, therefore:

$$r_{System\ I}{'} = r_{System\ II}{'} = 2 \left[\frac{batches}{unit\ time} \right]$$

Applying the reliability figures we determined, the expected system production rates are:

$$r_{System\ I} = \left(R_{System\ I} \right) \left(r_{System\ I}{'} \right) = (0.9807)\,(2) = 1.96 \left[\frac{batches}{unit\ time} \right]$$

$$r_{System\ II} = \left(R_{System\ II} \right) \left(r_{System\ II}{'} \right) = (0.7780)\,(2) = 1.55 \left[\frac{batches}{unit\ time} \right]$$

There are many factors in comparing functional and group technology layouts, such as scheduling, quality responsibility, and transportation. In these aspects, group technology is a superior form of layout, but from a reliability point of view only, the functional layout provides a higher reliability.

9.7 MORE COMPLEX CONFIGURATIONS

Typically most systems and, specifically, systems with a large number of components can represent a complex structure for reliability modeling and analysis. However, many parts of such systems are composed of sub-modules with components in series and parallel. These sub-modules can be reduced to single units using the simple series and pure parallel connection concepts. The units themselves can be further reduced if the configuration still allows. Therefore, for the simplified system representation, reliability calculations and analysis can be performed. Figure 9.11 illustrates one such system and its sequence of simplification.

At any level of the reduction phase, desired numerical values can be assigned and varied to see their effect on the reliability of the entire system.

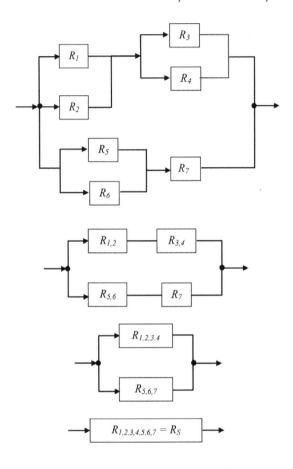

Figure 9.11. **Example of a reducible system.**

Systems whose interconnections cannot be simplified after certain stage into simple series, simple parallel, and combined series-parallel configurations and remain irreducible, or perhaps not desired to do so for the purpose of analysis, can be categorized as having a complex configuration. Figure 9.12 represents an already reduced version of a complex system. It is not possible to further simplify this representation.

There are several means and methods of handling and analyzing such systems. One method, which is readily amenable to using scientific or computational software, is the method of "path enumeration" and is described in Section 9.8.

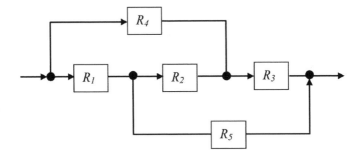

Figure 9.12. Simplest form of a complex reliability diagram.

9.8 THE METHOD OF PATH ENUMERATION

A complex configuration may look like the one shown in Figure 9.13.

Note that in this system, modules R_4 and R_5 can be simplified into one unit. But, for the sake of illustration, assume that it is desired that we maintain them as separate modules. For such a system, reliability is defined as the successful functioning of all the modules in at least one path. A path is a group of modules in series that form a connection between input and output when traversed in a stated direction. We also define the "minimal" path as a path that consists of a minimum number of modules. In other words, no module is traversed more than once in tracing the path.

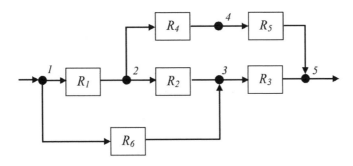

Figure 9.13. Reliability diagram of a complex system.

If P_i, $i = 1, 2,.., M$, are the minimal paths, then the reliability of the system R_s is expressed as:

$$R_s = Probability\ that\ at\ least\ one\ path\ is\ successful$$

or in mathematical terms:

$$R_S = Prob\left[\bigcup_{i=1}^{M} P_i\right]$$

where $\cup$ denotes the union operator.

Using the expansion rule for the probability of the union of M concurrent events, we have:

$$R_S = \sum_{i=1}^{M} Prob\left[P_i\right] - \sum_{i-1}^{M}\sum_{j>i}^{M} Prob\left[P_i \cap P_j\right] +$$

$$\sum_{i=1}^{M}\sum_{j>i}^{M}\sum_{K>j}^{M} Prob\left[P_i \cap P_j \cap P_k\right] + \dots + (-1)^{M-1} Prob\left[\bigcap_{i=1}^{M} P_i\right]$$

where $\cap$ denotes the intersection operator.

The system reliability R_s in terms of the individual module reliability can be restated as:

$$R_S = \sum_{i=1}^{M}\prod_{l \in P_i} R_l - \sum_{i=1}^{M}\sum_{j>i}^{M}\prod_{l \in P_i \cup P_j} R_l + \sum_{i=1}^{M}\sum_{j>i}^{M}\sum_{k>j}^{M}\prod_{l \in P_i \cup P_j \cup P_k} R_l + \dots + (-1)^{M-1}\prod_{l \in \cup_{i=1}^{M} P_i} R_l$$

It is actually unnecessary to work with these equations for any specific problem because a suitable algorithm is used instead, but it is useful to know its principles. As a check, the number of terms in this equation, and correspondingly the number of terms generated by the algorithm, is $(2^M - 1)$.

In the expression for R, the intersection terms must have the minimum of the common terms. For instance, if path P_1 has modules R_1, R_3, and R_4, and the path P_2 has the modules R_2, R_3, R_4, and R_5, the term $Prob\,[P_1 \cap P_2]$ should be expressed as $R_3 R_4$ and not as $R_3^2 R_4^2$.

Now using Figure 9.13 we present the path enumeration algorithm which generates the terms of the equation of the total system reliability, R_s.

Steps of the Procedure

1. Place nodes in between any two adjacent modules. This is not an essential step, but it can be useful in tracing large reliability graphs.

2. Carefully identify all minimal paths and display them in a table in node-trace and module-trace forms from input to output. There are three ($M = 3$) minimal paths in the system shown in Figure 9.13.

3. Find all the required union of the minimal paths.

4. Each minimal path or path-union is akin to a series configuration. Give each a reliability expression in terms of their module reliabilities.

5. Sum all the reliability expressions obtained according to the equation for R_s above, with the corresponding positive and negative signs.

The path-finding algorithm can also be applied in a tabular-visual form. This is demonstrated using Figure 9.13. The minimum paths are shown in Table 9.2.

Table 9.2. Trace of the minimum paths

Path	Trace (in Node Form)	Trace (in Module Form)
P_1	$1 \rightarrow 2 \rightarrow 3 \rightarrow 5$	$R_1 \rightarrow R_2 \rightarrow R_3$
P_2	$1 \rightarrow 3 \rightarrow 5$	$R_6 \rightarrow R_3$
P_3	$1 \rightarrow 2 \rightarrow 4 \rightarrow 5$	$R_1 \rightarrow R_4 \rightarrow R_4$

We express each path in a binary format. For example, for path P_1, the binary representation is given in Table 9.3. The boxes containing the digit 1 indicate the module is part of the path. It is not absolutely necessary, but it greatly facilitates generating the positive/negative alternating signs of the terms of the reliability equation, if we associate a negative sign to each minimal path at the beginning.

Table 9.3. Binary representation of a path

Path	Sign	Module					
		R_1	R_2	R_3	R_4	R_5	R_6
P_1	-	1	1	1	0	0	0

We now proceed to obtain the union of paths using a bitwise logical operation in the ascending order of paths and enumerate them. As the unions of the paths are determined, the sign is switched. For the entire system, the minimal paths, minimal path unions, and their binary representation are shown in Table 9.4.

Table 9.4. Binary representation of paths and path unions

Path and Path Unions	Sign	Module					
		R_1	R_2	R_3	R_4	R_5	R_6
P_1	-	1	1	1	0	0	0
P_2	-	0	0	1	0	0	1
P_3	-	1	0	0	1	1	0
P_1 or P_2	+	1	1	1	0	0	1
P_1 or P_3	+	1	1	1	1	1	0
P_2 or P_3	+	1	0	1	1	1	1
(P_1 or P_2) or P_3	-	1	1	1	1	1	1

One way of obtaining the correct sign is to assign a negative sign to the odd number of paths and path unions (P_1, P_2, P_3, and $P_1P_2P_3$), and a positive sign to the even number of path unions (P_1P_2, P_1P_3, and P_2P_3). Also, note that the number of rows in Table 9.3, which is the number of terms in the system reliability equation, is $Z = 2^M - 1 = 2^3 - 1 = 7$.

Next, for every module that has a "1" in the corresponding box, we form the products that correspond to the terms of the reliability equation. The sum of the results will give the system reliability with a negative sign. Therefore:

$$-R_S = -[R_1R_2R_3 + R_3R_6 + R_1R_4R_5] + [R_1R_2R_3R_6 + R_1R_2R_3R_4R_5 + R_1R_3R_4R_5R_6] - [R_1R_2R_3R_4R_5R_6]$$

By switching the signs of the terms, the system reliability is:

$$R_S = [R_1R_2R_3 + R_3R_6 + R_1R_4R_5] - [R_1R_2R_3R_6 + R_1R_2R_3R_4R_5 + R_1R_3R_4R_5R_6] + [R_1R_2R_3R_4R_5R_6]$$

Given numerical values R_i, $i = 1, \ldots 6$, the numerical value of the system reliability can be calculated.

9.8.1 Sensitivity Analysis

Determining a numerical value for the total system reliability tells us only how reliable the system might be. It does not by itself help us in getting an insight into the system's characteristics or improving it. Sensitivity analysis is an excellent extension to the path enumeration method in this respect.

A function is called bilinear if it is in the form of $f(x,y)$ and is linearly related to x and y. It is also possible to liken the equation for R_s to a bilinear function of module reliability. (Nowhere in the system reliability equation any of the R_i appear with power greater than 1, so R_s is a linear function of R_i.) The bilinear characteristic can be utilized by assuming any R_i at its extreme values, that is $R_s = 1$ and $R_i = 0$, to extract important information—namely, the sensitivity of the system reliability to module R_i. The module sensitivity s_{R_i} is determined as follows:

$$S_{R_i} = \frac{\partial R_S}{\partial R_i} = \frac{R_S\left(R_i = 1\right) - R_S\left(R_i = 0\right)}{1 - 0} = R_S\left(R_i = 1\right) - R_S\left(R_i = 0\right) i = 1,\dots N$$

N is total number of modules. Basically, this means that R_s is evaluated at *Max R_i* and at *Min R_i*, the difference being the sensitivity.

Example 9.4

If all the modules in the reliability diagram of Figure 9.13 have reliability of 0.9:

(a) Calculate the sensitivity of the system reliability with respect to modules R_1 and R_4.

(b) Explain the reason for the difference.

Solution

Part (a): We determined that the reliability expression for the system was:

$$R_S = \left[R_1R_2R_3 + R_3R_6 + R_1R_4R_5\right] - \left[R_1R_2R_3R_6 + R_1R_2R_3R_4R_5 + R_1R_3R_4R_5R_6\right] + \left[R_1R_2R_3R_4R_5R_6\right]$$

Module sensitivity was also determined as:

$$S_{R_i} = \frac{\partial R_S}{\partial R_i} = R_S\left(R_i = 1\right) - R_S\left(R_i = 0\right)$$

The sensitivity of the system reliability with respect to unit R_1 can be calculated using the numerical values:

$$R_2 = R_3 = R_4 = R_5 = R_5 = 0.9, R_1 = 1 \text{ and } R_1 = 0$$
$$S_{R_1} = R_S\left(R_1 = 1\right) - R_S\left(R_1 = 0\right) = 0.9878 - 0.81 = 0.1778$$

Similarly, the sensitivity of the system reliability with respect to unit R_4 can be calculated using the numerical values:

$$R_1 = R_2 = R_3 = R_5 = R_6 = 0.9, \text{ and } R_4 = 1 \text{ and } R_4 = 0$$

$$S_{R_4} = R_S(R_4 = 1) - R_S(R_4 = 0) = 0.9707 - 0.8829 = 0.0878$$

Part (b): We see that the sensitivity of the system reliability with respect to unit R_1 is about twice as much the same with respect to unit R_4. If we examine the system reliability diagram, Figure 9.13, we can identify three minimum paths. R_1 is a member of two paths, whereas R_4 belongs one path. Therefore, R_1 is twice as likely to affect the reliability of the system.

EXERCISES

9-1   What the definition of *reliability*?

9-2   How is the reliability of a component or a system expressed?

9-3   Name two functionally different systems that consist primarily of serially connected modules.

9-4   Name two functionally different systems that consist primarily of parallel connected modules.

9-5   A system consists of four units connected in series. The unit reliabilities are:

$$R_1 = 0.90, R_2 = 0.97, R_3 = 0.8 \text{ and } R_4 = 0.85$$

It is desired to increase the total system reliability by adding some parallel units. Space and cost considerations only allow adding either two identical parallel units to the existing unit R_3, or one identical unit to the existing unit R_4. Determine which option will be more effective.

9-6   For the system diagram given in Figure 9.14 use the path enumeration and sensitivity analysis to determine:

a)   The system reliability

b)   The sensitivity of the system reliability with respect to R_5 and R_6

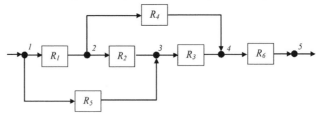

Figure 9.14.

10 QUEUEING THEORY

Chapter 10

10.1 INTRODUCTION

Many situations arise in manufacturing plants, industrial and governmental organizations, and other settings where queues form if the service required by a customer or entity is not immediately available. The instances of such demands are generally random in nature, and they are often caused by the engagement of the servers to attend to the existing customers in the system. Each demand normally requires a different duration of service. Therefore, the availability of service is also random and irregular in nature. These variabilities can cause queues to build up frequently and diminish over time.

The mathematical theory of queueing systems is rich, and its treatment is rather complex. But many different concepts have been worked out to fit a variety of real scenarios, and the developed set of equations provide insight and means to analyze and influence the behavior of the underlying system. The primary factors in using the queueing theory and obtaining reliable results are to have appropriate data and determine which model is best suited for the system being studied.

10.2 ELEMENTS OF QUEUEING SYSTEMS

Essentially, the queueing theory offers a modeling concept that is mathematical in nature and composed of a set of equations for specific problems. For modeling and analysis of queueing systems, the following key information is required:

217

- Arrival pattern of jobs, customers, or any entity of interest
- Queue discipline by which the entities join the system
- Configuration of service facilities for the flow of entities
- Service duration for each entity
- Metrics of analysis
- Options for system modification

In the following sections we describe these elements.

10.2.1 Arrival Pattern

For most queueing problems the arrival of entities is an independent random event beyond the direct control of the system's operator. If arrival data are available from past experience, or they have been recorded, appropriate probability distributions can be found to pseudo-generate the arrival times or the number of arrivals within a specific period of time for the mathematical modeling.

10.2.2 Queue Discipline

When customers or entities arrive and service is not available, they line up in a queue. Some systems may have multiple queues of equal importance, and the arriving entity or customer can choose the queue; however, under identical circumstances, it is highly likely that they join the shorter queue. The nature of the lineup is governed by certain rules known as "queue disciplines"; the following are typical examples:

- *First Come First Served (FCFS).* With this discipline, the arriving entities join the end or tail of the existing queue. This is the most common discipline that is easy to apply and follow in many practical situations. This rule is also called "First-In-First-Out."

- *Last Come First Served (LCFS).* With this discipline arriving entities go the beginning or head of the queue. This rule is less common, at least intentionally, but unavoidable for some systems. It applies, for example, when entities are stored in a limited-access warehouse or stacked vertically, where the last entity must be removed (attended to) first to access the previous other entities. This discipline is also called "Last-In-First-Out."

- *High-Value First (HVF).* In order to implement this discipline, a certain attribute of the arriving entities is used to rank them for the lineup. For example, for relevant reasons, the weight or volume of an entity may be used to assign priorities. An example would be the loading of cargo ships, where, for stability and efficient use of space, heavy or bulky items are often given high priority.

- *Low-Value First (LVF).* This is similar to the HVF discipline, only applied in reverse order. Similarly, a certain attribute of the arriving entities is used to prioritize them for service. For example, entities requiring short service time may be given high priority to reduce the queue length and have a larger number of service tasks accomplished in a shorter period of time.

10.2.3 Service Arrangement

The service facility can have single or multiple channels (multiple servers). This is often a design or controllable feature of a queueing system. Because the arrival patterns cannot always be altered at will by the analyst as a design factor, if long lineups and delays are observed, additional service channels can be added. A larger, more complex queueing system may have multiple or parallel points of arrival, and have many service channels in series. The modeling and analysis of such systems may pose a challenge for implementing the queueing concepts. Such a system can be readily analyzed using Discrete Event Simulation.

10.2.4 Service Duration

The service times depend on the nature of the service required and the capability of the service mechanism. The service times typically have much less variability than the arrival patterns or, more specifically, inter-arrival times. In some cases, the service time may even be constant.

Once the requirements of a queueing analysis have been determined, and information about the elements of the queueing system has been gathered, a model can be developed, and performance indicators, or metrics, can be computed.

10.2.5 Performance Metrics

The typical measures or metrics are the average number of customers or entities in the queue, and the average time an entity spends in the queue over an extended period of operation. More metrics are listed in Section 10.3.1, Table 10.1.

10.2.6 System Modification

The ultimate goal of the queueing modeling and analysis is to determine problem areas and investigate the feasibility and effects of various modifications. Three factors can be considered for change:

- *Arrival pattern.* This is often an independent variable for the system as a whole, but by adding new service channels in parallel, the arrival rate per channel will change. Another way of influencing an arrival pattern is to force some entities to "balk" (turn away) and not enter the system. Such as imposing a weight restriction for trucks entering a loading/unloading bay.

- *Service time.* Changing the mechanism of the service facility, such as replacing older equipment with faster and perhaps automated equipment, reduces the service time and, hence, the queue length.

- *Service facility configuration.* Depending on the problem situation, and the consideration given to the availability of space and other limitations, the service facility can be reconfigured. For example, parallel entrance and service channels can be added, or a service point can be split into faster serial processes.

In some cases, it is only required to change the numerical values of the parameters used, while in other situations where major changes are being considered, the queueing model must be reconstructed.

10.3 QUEUEING MODELS

Many forms of queueing models with different arrival patterns, service times, configuration, and operational conditions have been developed. Many real systems can be modeled using these standard forms. More complex systems can also be modeled by judiciously combining the standard models. Figure 10.1 shows several forms of queueing systems. In the following section, we review simple single-channel and multiple-channel models.

10.3.1 Single-Channel Queues

A number of assumptions that we explicitly did not mention earlier must be made in using the queueing theory. We consider a queueing system with the following characteristics:

- Infinite population of potential customers. This means that there will be no interruption in the probabilistic generation of the customers, such as interruption due to the night shift.
- FCFS queue discipline.
- Probabilistic service times.
- Single-channel/single-server arrangement.

If the number of arrivals and the number of services performed per given time follow the Poisson distribution, then the queueing system is referred to as a simple queue.

Single-line, single channel:

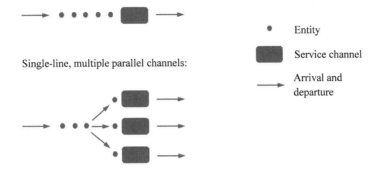

Single-line, multiple parallel channels:

Single-line, single channel multiple serial phases:

Single-line, multiple parallel channels, multiple serial phases:

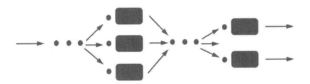

Figure 10.1. **Models of queueing systems.**

Figure 10.1 shows a number of single and multiple-channel queueing configurations. Over an extended period of time, when the system is assumed to have stabilized, some simple yet useful results about the average performance metrics have been mathematically derived. The single most important parameter of a simple single-channel queue is its "traffic intensity" or "utilization factor," defined as:

$$Traffic\ intensity = \frac{Mean\ arrival\ rate}{Mean\ service\ rate}$$

Using the common mathematical symbols:

$$\rho = \frac{\lambda}{\mu}$$

where also:

$$Mean\ arrival\ rate\ =\ \frac{1}{Mean\ inter\text{-}arrival\ time}$$

$$Mean\ service\ rate\ =\ \frac{1}{Mean\ service\ time}$$

It is necessary for the traffic intensity to be less than one (sometimes much less than one) for the system to stabilize in the long run. This means that the average arrival rate is less than the average service rate. Even with intuition it can be expected that otherwise the queue length will tend to grow indefinitely in the long run. Performance measures of a stabilized simple queue are given in Table 10.1.

Table 10.1. Metrics for single channel queue

Metric	Formula
Probability that the service facility is idle. That is, the probability of zero entity in the system.	$P_0 = \left(1 - \frac{\lambda}{\mu}\right) = (1-\rho)$
Probability of an arriving entity having to wait for service.	$P_w = \frac{\lambda}{\mu}$
Probability of n entities in the system.	$P_n = \left(\frac{\lambda}{\mu}\right)^n P_0 = \rho^n (1-\rho)$
Average number of entities waiting for service in the queue.	$\dfrac{\lambda^2}{\mu(\mu-\lambda)} \qquad \dfrac{\rho^2}{1-}$
Average number of entities in the system.	$N = N_q + \dfrac{\lambda}{\mu} = \dfrac{\rho}{1-\rho}$
Average time an entity waits for service in the queue.	$T_q = \dfrac{N_q}{\lambda} = \dfrac{\rho}{\mu(1-\rho)}$
Average time an entity is in the system.	$T = T_q + \dfrac{1}{\mu} = \dfrac{1}{\mu(1-\rho)}$

10.3.2 Multiple-Channel Queues

For simple systems, this is a common model. If not explicitly specified, this model has one queue for a number of service channels that are in parallel. Performance measures of a stabilized simple single-line c parallel channel system are given in Table 10.2. For a multiple-channel queue, for the system to be stable in the long run, the traffic intensity must also be less than unity, that is:

$$\rho = \frac{\lambda}{c\mu} < 1$$

Table 10.2. Metrics for multiple channel queue

Metric	Formula
Probability that all c service channels are idle. That is, the probability of zero entity in the system.	$P_0 = \dfrac{1}{\left[\displaystyle\sum_{n=0}^{c-1} \dfrac{\left(\frac{\lambda}{\mu}\right)^n}{n!}\right] + \left[\dfrac{\left(\frac{\lambda}{\mu}\right)^c}{c!}\right]\left(\dfrac{c\mu}{c\mu - \lambda}\right)}$
Probability of an arriving entity having to wait for service.	$P_w = \dfrac{1}{c!}\left(\dfrac{\lambda}{\mu}\right)^c \left(\dfrac{c\mu}{c\mu - \lambda}\right) P_0$
Probability of n entities in the system.	$P_n = \dfrac{\left(\frac{\lambda}{\mu}\right)^n}{c!\,c^{n-c}} P_0 \text{ for } n > c$ $P_n = \dfrac{\left(\frac{\lambda}{\mu}\right)^n}{n!} P_0 \text{ for } 0 \leq n \leq c$
Average number of entities waiting for service in the queue.	$N_q = \dfrac{c\lambda\mu\left(\frac{\lambda}{\mu}\right)^c}{c!\left(c\mu - \lambda\right)^2} P_0$
Average number of entities in the system.	$N = N_q + \dfrac{\lambda}{\mu}$
Average time an entity waits for service in the queue.	$T_q = \dfrac{N_q}{\lambda}$
Average time an entity is in the system.	$T = T_q + \dfrac{1}{\mu}$

Example 10.1

In an aircraft maintenance facility, there is one lineup of planes and three identical service hangars in parallel. Relevant data have been recorded, and the analyses show that both the arrival and service patterns follow a Poisson distribution. The data are given as:

- Arrival rate, $\lambda = 10$ *aricraft / month*
- Service rate, $\mu = 4$ *aircraft / month*
- Number of service hangers, $c = 3$

Determine:

a. All performance metrics (except P_n) given in Table 10.2.

b. The management is planning to increase the number of service hangars to four. Recompute all the performance metrics as in (a).

c. Compare and comment on the results of (a) and (b).

d. If the management contemplates implementing the new plan, what factors should they consider before doing so?

Solution

(a)

First, we check that the traffic intensity is acceptable:

$$\rho = \frac{\lambda}{c\mu} = \frac{10}{3\,(4)} = 0.833$$

Since , $\rho < 1$ we can use the multichannel queueing system equations:

$$P_0 = \frac{1}{\left[\sum_{n=0}^{c-1} \frac{\left(\frac{\lambda}{\mu} \right)^n}{n!} \right] + \left(\frac{\left(\frac{\lambda}{\mu} \right)^c}{c!} \right) \left(\frac{c\mu}{c\mu - \lambda} \right)} =$$

$$= \frac{1}{\left[\frac{\left(\frac{10}{4} \right)^0}{0!} + \frac{\left(\frac{10}{4} \right)^1}{1!} + \frac{\left(\frac{10}{4} \right)^2}{2!} \right] + \left(\frac{\left(\frac{10}{4} \right)^3}{3!} \right) \left[\frac{3(4)}{3(4) - 10} \right]} = 0.045$$

$$P_w = \frac{1}{c!} \left(\frac{\lambda}{\mu} \right)^c \left(\frac{c\mu}{c\mu - \lambda} \right) P_0 = \frac{1}{3!} \left(\frac{10}{4} \right)^3 \left[\frac{3(4)}{3(4) - 10} \right] (0.045) = 0.703$$

$$N_q = \frac{c\lambda\mu \left(\frac{\lambda}{\mu} \right)^c}{c! \, (c\mu - \lambda)^2} P_0 = \frac{3 \, (10) \, (4) \left(\frac{10}{4} \right)^3}{3! \, [3 \, (4) - 10]^2} = 3.5$$

$$N = N_q + \frac{\lambda}{\mu} = 3.5 + \frac{10}{4} = 6$$

$$T_q = \frac{N_q}{\lambda} = \frac{3.5}{10} = 0.35 \, [\, month \,] = 10.5 \, [\, days \,]$$

$$T = T_q + \frac{1}{\mu} = 0.35 + \frac{1}{4} = 0.6 \, [\, month \,] = 18 \, [\, days \,]$$

(b)

We repeat the same computation with $c = 4$:

$$\rho = \frac{\lambda}{c\mu} = \frac{10}{4 \, (4)} = 0.625$$

Since $\rho < 1$, we proceed to use the multichannel queueing system equations and present the numerical values as:

$$P_0 = 0.703$$

$$P_w = 0.31$$

$$N_q = 0.5$$

$$N = 3.0$$

$$T_q = 0.05 \, [\, month \,] = 1.5 \, [\, days \,]$$

$$T = 0.30 \left[month\right] = 9\left[days\right]$$

(c)

By increasing the number of service hangars from three to four, a change of 33 percent, significant improvements are achieved in all measures. Notably, the waiting time in the queue is reduced from 10.5 [days] to only 1.5 [days], and the time spent in the system is shortened by 50 percent.

(d)

The investment in an extra service hangar will certainly be substantial. On the other hand, time reduced in waiting can be used for flying. The management must, on the basis of the number of aircraft serviced per year, and the time saved per aircraft, determine the potential annual income from additional flying time, compare it with the cost investment capital required, and make the appropriate decision.

EXERCISES

10-1 In using the queueing theory, why should the traffic intensity (utilization factor) be less than unity?

10-2 Assume a single-channel queueing system with arrival rate λ and service rate μ. Then assume another queueing system with arrival rate of 2λ, service rate μ, and two channels in parallel. Select and determine some performance metrics and compare the two systems. Comment on your observations.

10-3 In a similar fashion to Exercise 10-2, using all the metrics in Table 10.1 compare two queueing systems: one with arrival and service rates of λ and μ, and the other with arrival and service rates of 2λ and 2μ.

10-4 Cargo ships arrive at a port according to a negative exponential distribution of inter-arrival time, at a rate of one every five hours. The time a ship spends at a dock for unloading and loading also has a negative exponential distribution, and its average is eighteen hours. The port operates twenty-four hours a day, and there is a plan to increase the number of docks to keep average waiting time below nine hours. How many docks in total should the port have to accomplish this?

Chapter **11** APPLICATION OF **PRINCIPLES**

11.1 INTRODUCTION

In the previous chapters, we addressed some fundamental concepts and a fair range of techniques that have proven useful in the field of industrial engineering.

The concepts covered in the previous chapters can be applied individually in many manufacturing and service industries. Every concept has reducing cost and inefficiencies as its goal, and, consequently, ensuring a better return on investment of capital, time, or effort.

It makes sense, however, to be in a position to apply as many ideas as possible from the preceding chapters, bearing in mind that no two organizations are exactly alike. They have different sizes, missions, styles of management, and unique facilities and personnel; thus, any treatment must fit the goals of the organization. In this last chapter, we consider how we can make collective use of these techniques to impact the entire organization on a large scale.

The first step is to master the fundamentals of industrial engineering and gain experience in their application in various settings on modest problems. The next step is the study of advanced and dedicated books on subjects of interest and the handling of larger problem cases and projects.

In putting it all together, there are several industrial engineering, operations management, and controls techniques available. These techniques work at the macro-level and encompass the organization in its entirety. They aim to improve the quality of the products and services and to increase efficiency and reduce waste. Four of these methods are the following:

- 5S concept
- Supply-chain management
- Six-Sigma technique
- Lean thinking

11.2 5S CONCEPT

5S is a workplace organization idea that originates from five Japanese words: *seiri, seiton, seiso, seiketsu,* and *shitsuke.* These words have been translated into English as *sort, set-in-order (or straighten), shine (or sweep), standardize,* and *sustain.* Each word relates to a group of tasks that describes manners of workplace organization to improve efficiency and effectiveness of facilities, equipment, and personnel. To make their relevance clearer and before detailed explanations are given, we present an overview of the 5S concept in Table 11.1.

Table 11.1. Brief description of the 5S concept

Task	Focus and Explanation
Sort	Removing of unneeded materials and objects
Set-in-order	Organizing physical facilities and material for easy access and movement of equipment and personnel
Shine	Maintaining an orderly and clean workplace by inspecting on daily basis
Standardize	Implementing standards for all activities, and regularly reviewing and improving the conditions brought about by the first three S-tasks
Sustain	Keeping to the rules and conditions set in the first four S-tasks to maintain the standards and continue to improve every day

11.2.1 Detailed Explanation of 5S Action Items

All 5S tasks are important, but *sort* is the most essential, and it has a significant impact on the work environment. The following are the types of actions that must be undertaken in this step:

- Inspect the workplace and remove items that are not immediately needed.
- Make work easier by eliminating obstacles.
- Prevent hazards and accidents by removing unnecessary and dangerous distractions.

- Remove all waste material and scraps.
- Improve and maintain the tidiness achieved.

In essence, *sort* provides a safe work environment and makes it easier to locate the items that may be needed or be moved around the plant.

Set-in-order or *straighten* is a complement to *sort*. In fact, *sort* is more about clearing obstacles, and *set-in-order* is about organizing the remaining necessary materials and equipment. The following is a list of typical action items:

- Arrange all necessary items so that they can be easily accessed when needed.
- Arrange work cells and workstation such that tooling and supplies are in close proximity for easy access.
- Investigate design and tooling changes to improve efficiency.
- Establish a proper priority for accessing and borrowing tools. Often, First-Come-First-Served (FCFS) is a good basis, but it is not a universal rule.
- Make work flow smooth and easy.
- Organize physical facilities for convenient and safe movement of equipment, personnel, and material.
- Improve and maintain the order accomplished.

Performing the *set-in-order* task reduces transportation times and increases temporary storage space for items in transit.

Every S-task should be inspected, improved, and maintained to benefit from the efforts spent. Relevant action items for the *shine* or *sweep* task include the following:

- Clean the personal and group workplaces thoroughly.
- Use the process of cleaning as a checkup and safety tool.
- Plan proper maintenance to prevent machinery and equipment deterioration.
- Maintain the workplace as clean, pleasing, and safe.

This activity contributes to professional and pleasant work conditions.

The conditions accomplished in the first three S-tasks can be greatly enhanced by the *standardize* process. It will also add professional ambiance and significantly increase efficiency. Some action items include the following:

- Standardize and implement the best practices in the work area.
- Maintain every item, system or procedure according to its standard of use.
- Return every item to its right location after each use.
- Subject everything to standard and uniformly apply it to similar procedures.

Once this task accomplished, the benefits are in the better consistency of all processes and activities, and the quality of output.

Sustain advises abiding by the rules so as to maintain the status quo and standards achieved and to continue to improve. Specifically, *sustain* calls for the following actions:

- Keep systems and operations in proper working order.
- Perform regular audits of systems and operations.
- Use training and encourage discipline to sustain and improve the achieved results.

This last step in the 5S concept prevents deterioration of the state of the improved system.

Implementing the components of the 5S concept can be done by one individual depending on the size of the enterprise. However, it is always desirable that a small core team undertakes the entire implementation process, and along the way seeks input and further ideas from the personnel involved.

11.3 SUPPLY-CHAIN MANAGEMENT

Supply-chain management is the process of planning for a well-coordinated flow of goods, services, and information relating to six core components of the supplier, manufacturer, distributor, logistics, retailer, and customer. Depending on the organization and the nature of its activities, the chain or, more generally, the link or network can be represented in a variety of forms. Each component can also receive different emphasis and importance for the same reasons. Figure 11.1 shows a generic supply chain, and Figure 11.2 shows an example of a customer-centered supply network. In fact, a supply chain can be thought of as a special case of a supply network; however, it has become customary to refer to all forms of relationship and interaction between the facilities and functions as a "supply chain."

11.3.1 Principles of Supply-Chain Management

Supply-chain management can be improved and coordinated in many ways. Its successful application in the manufacturing industry has led to the development of seven effective principles. The essence of these principles is given in definitions in the form of instructions.

Segment customers based on their needs. Not all customers are alike. They differ in terms of the products and services they need, and they differ in the level of

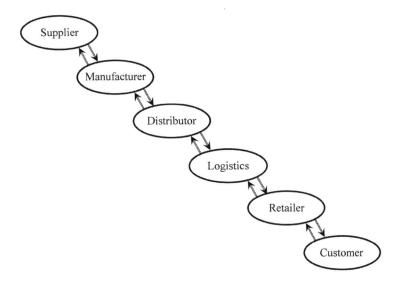

Figure 11.1. Generic supply chain.

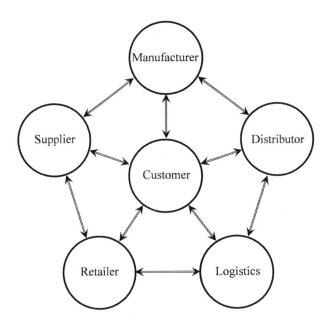

Figure 11.2. A customer-centered supply network.

quality they desire and are willing to pay for. Segmenting customers into distinct groups can help identify these needs better and examine whether cost-effective changes can be made to the products or services.

Customize the logistics network. Once the first principle has been applied and the customer groups have been identified, customize the logistics network, such as inventory, warehousing, customer relations, and transportation to service each group's requirements.

Synchronize operations to demand. Pay attention to market signals. Use forecasting to coordinate demand with material requirements, resources allocation, and scheduling of operations.

Differentiate the product closer to the customer. This principle means that one should keep the requirements, in terms of inventory, processes, and other necessary tasks, as common as possible, and only near the end and closer to the customer differentiate (categorize) the product and services in the supply chain. This shortens the lead time and expedites the delivery to customers.

Strategically manage the source of supply. Manage sources of supplies strategically to reduce the cost of acquiring materials and services. The cheapest or closest sources of supply may not always be the best. Consistency and reliability in the sources of supply are essential to quality, on-time delivery, and customer satisfaction. In the long run, these translate to reduced costs of procurement.

Develop a supply-chain-wide technology strategy. Provide multiple levels of decision making, giving a clear view of the flow of products, services, and information. The technology strategy should encompass information technology and other technologies to facilitate day-to-day and long-term activities along the supply chain.

Use supply-chain-spanning performance measures. Develop through-the-chain end-to-end performance indicators to measure effectiveness and efficiency. Use the metrics to improve and control the processes.

11.4 SIX-SIGMA TECHNIQUES

Six-Sigma is a management philosophy composed of a set of techniques for process improvement. Essentially, Six-Sigma aims to improve the quality of the process output by identifying and eliminating the causes of defects, thereby minimizing the variability of the process. Six-Sigma is used in diverse manufacturing and service industries. It is based on quality management and statistical methods. Its implementation requires a team of experts in the field who work on a specifically defined project and follow a defined set of steps to achieve the desired goal. Typical goals include, for example, reduced costs, reduced cycle time, or improved customer satisfaction—all through improving the quality characteristics of the process and its output.

A Six-Sigma process is designed such that 99.99966 percent of all observations, occurrences, or "opportunities" should be defect-free, the complement of which will be 0.00034 percent or 3.4 defective occurrences per million opportunities. Recalling the concept of three-sigma limits, used in the discussion of the statistical quality control in Chapter 7, it is clear that Six-Sigma calls for setting extremely rigorous performance objectives. This relates to collecting data, analyzing the results, and developing and implementing a plan to reduce the defects in the process output.

Two methodologies have been defined for two areas of application, and, correspondingly, two acronyms have also been defined:

1. DMAIC (pronounced "duh-may-ick") is used for projects aimed at improving an existing process.

2. DMADV (pronounced "duh-mad-vee") is used for projects aimed at designing a new product or process.

DMAIC stands for the following five phases:

1. *Define* the system, the project goals, and the customer requirements.

2. *Measure* the current key process, collect relevant performance data, and calculate the process capability.

3. *Analyze* the data to investigate cause-and-effect relationships. From the relationships, determine the root causes of the defect or the error.

4. *Improve* the process or fix the error based on the data analysis using techniques such as the design of experiments, simulation, brainstorming, and pilot runs.

5. *Control* the future process performance and ensure that any deviations from the target are corrected in a timely fashion. Implement statistical process control and control charts, develop standards, and continuously monitor the process to stay on course.

DMADV also has five phases and stands for the following:

1. *Define* design goals consistent with the customer needs and the management strategy.

2. *Measure* and identify all characteristics that are critical to quality (CTQs), measure product and production process capabilities, and assess risks.

3. *Analyze* products and processes, and make adjustments for improvements.

4. *Design* an improved alternative, by assessing the enhanced process, and seeking feedback from test-based-customers.

5. *Verify* the design as final, implement the process, and continue monitoring the product and customer feedback.

According to the Motorola company, who initiated the Six-Sigma methodology, the standard Six-Sigma processes are executed by professionals designated as Six-Sigma Green Belts or Six-Sigma Black Belts, whose work is then overseen by Six-Sigma Master Black Belts to ensure total compliance with the concepts.

It has been reported that by implementing Six-Sigma technique, up to 50 percent process cost reduction, cycle-time improvement, and less waste of materials can be achieved. Adopting and committing to Six-Sigma can be costly and may take several years before a company begins to realize the benefits.

11.5 LEAN THINKING

The emphasis in Six-Sigma is on eliminating defects and reducing variability in processes and their output. Lean thinking is a similar concept, but it focuses on eliminating waste and coordinating activities from end to end. This will lead to optimality in all aspects of the operation of an enterprise. It was primarily intended for the automotive industry as "lean manufacturing," but the concept has been expanded in its scope of application.

Lean Thinking

Some variants of lean thinking include the following:

-   Lean manufacturing
-   Lean Sigma
-   Lean supply chain

In fact, lean thinking can be applied in any enterprise, when its fundamental concepts have been understood. With reference to lean manufacturing, the core concept of lean thinking is based on three Japanese words.

In English *muda* means *waste*. As a process in the *lean* concept, it means the elimination of waste and loss. Excess inventory of parts with no added-value, unnecessarily occupied space, and avoidable costs are all prime examples of resources wasted.

Seven key resources that can benefit from *muda* have been identified as the following:

1. Transportation: moving products that are not required in the processing

2. Inventory: raw materials, work-in-progress, and finished products not being processed

3. Motion: unnecessary moving of equipment and walking of personnel

4. Waiting: idle and waiting for the next processing step.

5. Overproduction: producing excessive product inventory ahead of demand

6. Overprocessing: a consequence of improper design and process planning

7. Defects: the effort needed to inspect for defect and apply the corrections

In English *muri* means *overburden*. As a *lean* concept activity, it means balancing the production activities and arriving at the most appropriate production pace.

In English *mura* means *irregularity* or *variability*. In *lean* terms, it means the elimination of unevenness by balancing machine capacity and automating and task sharing by personnel.

Indeed, to fully benefit from the application of any of the 5S, supply-chain management, Six-Sigma, and lean thinking philosophies, management must be committed to providing facilities and a certain level of freedom, capital, and personnel. And the personnel charged with implementing the techniques must bear in mind the duties we outlined at the beginning of the book in Section 1.2.

EXERCISES

11-1   Using the concept of 5S, prepare a list of action items in your personal work environment.

11-2   As a follow-up to Exercise 11-1, develop a list of action items for your section in your work place.

11-3   Identify by name the six elements of the supply chain in your place of work. Then, by properly connecting the elements, determine whether they form a straight supply chain or a supply network.

11-4   In extension to Exercise 11-3, investigate the applicability of the seven principles of supply-chain management.

11-5   With reference to Six-Sigma, explore the applicability of DMAIC and DMADV in your workplace.

11-6   Reexamine Exercises 11-1 to 11-5 for a hypothetical or existing company of your choice.

11-7   In any of the above exercises, do you notice any potential gains? If so, make a list and provide a few action items for each.

Appendix **A**

EFFECTIVE
MEETINGS[*]

Between 35 and 50 percent of any manager's time is spent in meetings of one kind or another; many managers would put it even higher than this. Yet the conduct and outcome of meetings are often criticized severely. It is not uncommon to hear comments such as: "A committee is a body of people who keep minutes and waste time"; and "A meeting is a process by which a body of people decide when to have the next meeting." There is some truth in these comments and many valuable manpower hours can be wasted by ineffectual meetings, both formal and informal. Too often, this is because the person officially in charge of the meeting is ill prepared, lacks the expertise to lead and control a group, or is unaware of the communication problems that may arise. It may also be because those who attend the meeting have been briefed inadequately about the intended content. In this case, they, too, will lack preparation or arrive with hidden agendas—objectives of their own which have little to do with the original aims of the meeting. These will often dominate the discussion. The purpose of the material presented here is to improve the effectiveness of your meetings; saving you time, effort, money, and a great deal of frustration.

---

[*] "Effective Meetings" has been reproduced with the kind permissions of its author, Dr. Ghulam Shabir, managing director of DGS Pharma Consulting Ltd. and the Institute of Manufacturing, UK. The original manuscript appeared in volume 3, number 4 (2002) issue of the institute's *International Journal of Manufacturing Management*. The manuscript has been revised to fit the tone and language of the book.

WHAT IS A MEETING?

A meeting is an occasion when a group of people comes together to share ideas and experiences. Meetings widely differ in size, composition, organization, and purpose, but they all have certain things in common. They should all do the following:

- Exist for a purpose.
- Involve people communicating with one another.
- Have someone in control.

It is fundamental to the success of any meeting that its purpose is clear and relevant. Unless the meeting serves a useful purpose, it not only wastes time and effort but it can cause frustration, misunderstanding, and generally demotivate people. Meetings can become an end in themselves. This is particularly the case of regular meetings, which tend to take place because they always have. Regular meetings do have uses, but it is essential that there is some worthwhile business to discuss.

INFORMAL AND FORMAL MEETINGS

Differences

Informal meetings do not have a constitution, a set procedure, or specific needs for documentation; they are usually called and controlled by the most senior member present, in order to meet the requirements of routine planning or problem solving. They can be advisory in nature—the senior members retain the right to make the final decision. The majority of meetings held at work are of this type. Yet far too little consideration is given to the methods by which they are conducted and to their results. The person controlling the meeting has to achieve two basic objectives: to accomplish a task and to maintain the group as a team working toward this. "Informal" does not mean "unstructured."

Formal meetings always have a set constitution and procedure, as well as basic terms of reference. These are defined by memoranda, articles of association, acts of parliament, standing orders, or other forms of written constitution, according to the type of institution, public body, or organization in which they take place and the requirements of company law. Formal meetings have officers to conduct their business and record their proceedings and are controlled by a chairperson or president, who is elected or appointed and has well-defined duties. Meetings can be executive in nature, and decisions are taken by voting. Written documentation is an essential requirement. As the procedure is laid down, such meetings are perhaps a little easier to control but the main objectives, as for informal meetings, are just as relevant. Having contrasted the two types of meetings, it is worth seeing what they have in common.

Similarities

The similarities between these two broad divisions lie in the general objectives and aims and in the problems of leadership and control, which are common to both. Meetings serve a variety of purposes but basically they should be concerned with the following:

- Giving information or instruction
- Gathering information or opinions
- Persuading
- Planning
- Problem solving
- Decision making
- Initiating action

The way they are organized and the role of the leader of the meeting vary accordingly.

Categories of Informal Meetings

Informal meetings is the broad term covering a number of categories and these can be defined as follows:

- *The briefing meeting.* This is usually called by a manager in order to give instruction to the work group, to train them in new tasks, or to plan work. It is a highly useful method of communication, giving people an opportunity to interpret policy, test opinion, and provide personal explanations. It allows people to ask questions and clarify for themselves what is required. It may even result in new points and problems being raised, which the manager may not have considered, thus avoiding difficulties in advance.

- *The progress meeting.* This meeting reviews the progress of work, in discussing the problems that have arisen, measuring achievement, and planning further action. The advantages are similar to those of the briefing meeting. In addition, it allows problems to be detected and discussed at regular intervals so that they can be remedied quickly.

- *The planning meeting.* Similar to the previous two types, this meeting may initiate a project, plan its strategy, and allocate work to the appropriate people.

Its main purpose is to achieve a rational approach to the task, in consultation with all those who can offer expertise or who will take part in its completion, thus taking advantage of their advice and achieving their support.

- *The problem-solving meeting.* This is usually called to deal with a particular emergency or solve an important problem that has arisen. All those who can offer advice, evidence, help, or expertise should be consulted in order to get adequate information to reach a solution.

- *The brainstorming meeting (also called a "buzz group").* Connected with the two preceding categories, this is a particular approach that can be taken when planning (for example, the marketing strategy of a new product) or problem solving, when the causes of the problem may be particularly diverse, complex, or impenetrable. It is the most unstructured of the meetings, as anyone present has a right to call out any ideas, which may give a "lead." These are usually recorded on a flip chart or board. The elimination of the impossible or unlikely ideas is then followed up by an open discussion of the others until a few possible alternatives are chosen for further consideration and development.

Categories of Formal Meetings

This category in industry and commerce covers such things as annual general meetings and board meetings, which we will not discuss here.

THE RESPONSIBILITIES OF THE MEETING LEADER

The leader of any meeting has a variety of responsibilities, which include the following:

- Defining the purpose of the meeting
- Planning and preparation
- Conducting the meeting efficiently
- Controlling the discussion without doing all the talking
- Dealing effectively with problem situations and individuals
- Ensuring that the meeting keeps to the schedule
- Making sure everyone participates in the meeting
- Making sure the purpose of the meeting is achieved, and the members know what is expected of them as a result
- Liaising with the secretary, if there is one, to ensure that an efficient recording mechanism exists when required

In many instances, the secretary of the committee or meeting will take responsibility for organizing the date, time, and place of the meeting. The secretary may also draw up and circulate the agenda and write and distribute minutes. For maximum efficiency, there needs to be some level of consultation between the meeting leader and secretary about the agenda and minutes.

PLANNING AND PREPARATION

A well-planned meeting has much greater chance of success than the one that is called with five minutes' notice and has not been well thought out. There are four areas where planning and thorough preparation can assist the leader of the meeting:

1. Defining the purpose of the meeting
2. Planning the content of the meeting—the "agenda"
3. Deciding who should attend and for how long
4. Planning the domestic arrangement for the meeting

Defining the Purpose of the Meeting

Two questions need to be answered: What is the purpose of holding this meeting? Is the purpose best achieved by holding a meeting?

Many meetings are ineffective because the purpose is not clear. This applies especially to regular monthly meetings, for example. The main purpose of these sometimes seems to be the self-perpetuation of the monthly meetings.

It is, therefore, crucial that the person calling the meeting asks the question "what is the purpose" before doing anything else. It is sometimes a good idea to write the purpose down in one or two sentences so that it is clear and can be stated at the beginning of the meeting, if necessary. Once the purpose is defined, the second question should be asked. Sometimes the assumption is automatically made that the meeting is necessary when, in fact, several telephone calls, messages or a memo would serve as well. If the leader cannot clearly state the purpose of the meeting being called, a meeting is usually not necessary at all.

Planning the Content of the Meeting—the "Agenda"

For any meeting to succeed, the participants need to have a good idea of what is going to be discussed, and why they are there. The most effective way of ensuring this is to distribute an agenda detailing the topics to be discussed, who is to be there, where the meeting is to be held, and, if possible, the approximate time the meeting will take. In the case of a regular meeting the participants, timing and venue will normally remain the same and need not be highlighted each time. The agenda should fulfill two basic functions:

1. To inform the members of the topics to be dealt with, why they are being discussed, and what contribution is required
2. To act as a structural basis for discussion

The agenda should be as detailed as necessary and not just a list of headings. Checklist 1 details the stages necessary in the preparation of an agenda. This may be done by the secretary, but it is essential that the meeting leader is aware of the function and purpose of the agenda at this stage. The meeting leader also needs to plan a rough guide of how the meeting will be conducted. The actual skills of chairing will be important here and are detailed in a later section.

Planning the Arrangement for the Meeting

The leader is responsible for calling the meeting and deciding who is to attend, although, in some instances, the secretary will carry out this duty, together with booking the room and the time. Three points are worth mentioning here.

1   Recording the Meeting

In all cases, it is essential that someone be appointed to take notes or minutes. One of the main causes of unsuccessful meetings is the decision of the meeting leader to take notes. Neither role can be carried out effectively by the same person. Conflicts of interest may arise if the note taker is also a participating member of the meeting. Ideally, someone who is fully competent with the subject matter but not directly involved is in the best position to take notes. Where this is not possible, the note taking could be organized on a rotational basis, among those attending.

2   The Meeting Place

The meeting place should be comfortable and convenient and as free from noise and interruptions, such as telephone calls and "through traffic" of other people as possible.

3   Cooperating with the Secretary

The meeting leader and the secretary (or note/minute taker) can benefit from having a short get-together before the meeting starts. The following topics can be discussed:

*   *Technical terminology:* if the secretary is unfamiliar with any of the terminology used, the meeting leader can explain some of the words and terms most likely to arise.

*   *Structure of the meeting:* how the meeting will be chaired.

- *Type of the minute needed:* they can decide on the type of minutes needed, that is, action notes, verbatim record, threads of discussion and decisions, and so on.

- *Summarizing:* the secretary can encourage the use of summaries to make the note taking easier.

- *Clarifying the decision:* the secretary can establish the acceptability of interrupting the meeting for clarification if the secretary is unsure of what to note down for minutes.

Cooperation between both parties is essential to the success of meetings and will prevent lengthy drafting and redrafting of minutes afterward.

THE SKILLS OF CHAIRING MEETINGS

The leader of a meeting should have a number of skills to help maintain control of the discussion during meetings. The leader should not do all the talking and should ensure that everyone participates and no particular members dominate.

Opening the Meeting

The leader should open the meeting in a friendly but businesslike manner. This will help create the right climate. The meeting should start with an introductory statement summarizing what the meeting is for, what is known, what is required, and how it is going to be tackled.

Conducting the Meeting

In terms of leadership and control, there are two sets of specific objectives for the leader. The first set relates to the achievement of the task itself; the second refers to the control of the group working on that task. In each case, there are various functions, which the leader should perform. These can be listed as follows.

Task Functions

These keep the group working on the task or project and include the following:

(a) Initiating—proposing tasks and goals, defining the group problem and suggesting procedures or ideas

(b) Information/opinion-seeking—requesting facts, seeking relevant information and asking for suggestions or ideas

(c) Information/opinion-giving—stating a belief and providing relevant information

(d) Summarizing—pulling together related ideas, restating suggestions after the group discussion, and offering a decision or conclusion for the group

(e) Clarifying—elaborating, interpreting, or reflecting ideas, restating suggestion, clearing up any confusion, indicating alternatives and issues, and giving examples

(f) Consensus testing—checking to see how much agreement has been reached

(g) Action planning—delegating the tasks, now agreed, to those appropriate, and confirming them in writing

Human Relations or Group Maintenance Functions

These ensure that the group works as a team to achieve the objectives. They include the following:

(a) Encouragement—being friendly and responsive to the group, accepting contributions, and giving opportunities for recognition

(b) Expressing group feelings—sensing feelings, moods, and relationships within the group and sharing one's feelings with others

(c) Harmonizing—attempting to reconcile arguments and to reduce tensions, and getting people to explore their differences (not always an easy task)

(d) Modifying—staying conflicts-free, by adjusting stand to be able to admit error and maintain self-discipline for group cohesion (this function is particularly difficult for some)

(e) Keeping the channels of communication open—ensuring participation by providing procedures for discussion

(f) Evaluating—expressing standards for the group to achieve, evaluating the achievement, and the degree of commitment to action

The leader should be aware of the needs of the group and should draw out silent members and control the talkative. The leader should prevent private discussions in splinter groups. The leader's participation will vary according to the type of meeting, but it should be remembered that the maximum involvement of the leader means minimum participation by the members

Conducting the Meeting

Achievement of the objective is the whole purpose of the meeting. The leader needs to establish and emphasize this achievement and commend individuals and group contributions. Even if the result has not been entirely satisfactory, the leader should try to emphasize the positive aspect. The leader should make a final summary confirming the conclusions of the meeting and highlighting action and who is to take it.

Summaries

The use of interim and final summaries is one of the most powerful skills available to the leader.

Interim summaries serve to accomplish the following:

- Indicate the progress or lack of it.
- Refocus off-track discussions.
- Tie up one point and move to the next issue.
- Highlight important points.
- Guide the minute taker.
- Clarify misunderstandings.

The *final summary* establishes the group's conclusions, points to action, and gives members a sense of what has been achieved. Both interim and final summaries should be presented to the members for their agreement. This practice will also help the minute taker and cut down on unnecessary argument about the accuracy of the minutes.

PROBLEM SITUATIONS

The meeting leader may also have to deal with difficult situations in meetings, not only because of the subject matter, but also the following issues.

Unpopular Decisions

Leaders may find themselves under attack from individuals or the whole group. They must be prepared to present their case, acknowledging the weaknesses, anticipating the opponent's objections, and emphasizing the positive aspects. If there are members of the group in support of the case, encourage them to express their views—they may carry more weight than those of the leader.

Bad News

The leader must try to find a positive side.

Lack of Interest

The leader must recognize that every item on the agenda cannot be of equal interest to everyone. The likely causes of this are the following:

- The wrong items have been selected.
- Too many items are being dealt with, so attention wanes.

- The case is poorly presented so that relevance to members is not clear.

Some remedies are the following:

- Direct questions to individuals, for example, "How does this affect you?"
- Drop the subject entirely and return to it later, dealing with a more interesting subject in the meantime.

Discussion Going off Track

This may be indicative of a lack of interest. The group may start talking about other things or asking irrelevant questions. If there is genuine concern over something else, which is worrying the group, it is best to deal with it. Otherwise, frequent summarizing will help keep the discussion on course.

Status Differences

It is the leader's responsibility to protect the weaker members of the group. If there are more junior members present, it is often a good idea to ask for their thoughts first. They may be unwilling to express a view once more senior members have made their position clear.

People Problems

Perhaps the most difficult problem of leadership and control is one that concerns human relations—the individuals in the group have different personalities and objectives, which conditions how they behave. Although it is out of the scope this book to discuss the psychology of individuals and group dynamics, it may be worth giving some general principles:

- Every group contains those who wish to talk and those who, for a variety of reasons, are silent. It is a mistake to suppose that the extrovert talkers have necessarily the greatest contribution to make. Those who listen may have learned more and may have considerable expertise.
- In order to involve those who are silent, one or two strategies can be used. A procedure should be developed by which each person is asked for a contribution in turn, or the speaking time of those who normally have a great deal to say should be limited by inviting someone else to comment. Those members who are shy (or withdrawn for some other reason) should be asked questions about their particular area of expertise. This combines the advantage of asking them something, which they find easy to answer, thus developing their self-confidence and giving them special recognition, to which they are likely to respond.
- If a conflict develops between two members of a group, the rest of the group should be asked to comment on the two sides. This avoids driving the

conflict underground, where it is likely to smolder, leading to withdrawal and frustration. This is sometimes known as "the conference method."

- If aggression is shown toward the leader, the leader must paraphrase the aggressive statement in a milder form and ask the group to comment. If it becomes evident that there is truth in what is said, the leader must be prepared to modify his or her position. This should not be considered "backing down" but as a step that allows progress toward a new position.

- If it becomes evident that some members of the group are repeating themselves and holding on tenaciously to one view only, the discussion should be stopped and the alternative views expressed should be summarized. Often it is helpful to do this in writing if a visual aid is available (a flip chart or board in the room is useful). This prevents those members from retaining a mental image dominated by their view and can be used to move the discussion forward.

- When agreement seems to be present, a true consensus must be reached before defining what actions should be taken and allocating tasks. Otherwise people will depart from the meeting still in disagreement and proceed to sabotage the recommended action by expressing doubts to others—often their own subordinates.

- The leader must ensure that everyone knows what action is expected and that the meeting is not closed until this is decided. Too many meetings appear ineffective to those who have attended them because they leave unsure of what has been achieved and what results are expected. There is more motivation and satisfaction if they are certain on these fronts. A written reminder (action notes or minutes) may help.

STATING A CASE AT A MEETING

The best preparation and planning can be lost if someone finds it difficult to express their case coherently. Some people find it particularly difficult to put over their ideas to a group and will remain silent for fear of being ridiculed if they speak. Effective speakers know exactly what they want to say, have something to say, and say it well. In contrast, ineffective speakers are never clear in their own minds as to what they intend to say. As they speak, they move about in verbal confusion, following no set plan and developing no relevant argument.

Some of these problems can be overcome by following the guidelines below:

- Be sure of the facts. Sum them up briefly, and define any terms that need explaining. State the proposition.

- Face the facts. Weigh what is against you and anticipate objections. This also acts as a check on the soundness of your reasoning.

- Prove your case by selecting and highlighting the best reasons for the proposition. The strength of the argument will depend on the quality, not the quantity of your reasons.

- Show solid evidence. Have examples available to support the fact, but be careful not to slant the evidence to suit the case.

- End by repeating the proposition.

By following these simple rules and by having thought out the arguments before the meeting, persuading people to support your viewpoint will be made simpler.

MINUTES AND FURTHER FOLLOW-UP ACTION

It is essential that minutes are produced, but they will vary depending on the type of meeting. Formal committees sometimes require detailed minutes, whereas informal meetings may produce only an action list of things to be done as a result of the meeting. Whichever is the case, it is important that minutes do the following:

- Are produced quickly

- Are accurate

- Are not a verbatim report

- Show what action is required and by whom

Checklist 2 provides guidance on taking notes and writing up the minutes. The leader of a meeting should note the following points, which will help the minute taker to produce accurate and concise minutes:

- Summarize at the end of each point on the agenda. In most instances this is all that is required in the minutes, and a summary will help the minute taker, especially if the material is unfamiliar.

- Explain any unfamiliar terms or expressions, particularly technical jargon, which may be used during the meeting.

- Encourage the secretary to interrupt if a point has not been fully understood.

Where follow-up action is required, it is important to monitor progress. Giving individuals deadlines for follow-up action is useful. "As soon as possible" means something different to different people. If precise dates are given during the meeting and recorded in the minutes, monitoring progress is a much more straightforward task.

CHECKLIST 1: THE AGENDA

Subject Matter

Decide what topics need to be included on the agenda. It is always worthwhile, especially in the case of regular meetings, to check with those attending to see if they have any items they wish to include.

Avoid Being Vague

Do not have a vague heading "matters arising from the minutes." If matters need to be discussed against, place them on the agenda as specific items. The term *any other business* is often the cause of uncontrolled and time-wasting discussion. If members are asked beforehand whether they have items for inclusion on the agenda, a formal "Any Other Business" can quite easily be left out. This does not prevent last-minute items from being discussed, but it will prevent long, rambling, and unstructured discussions from taking place.

Be Logical

Put the items in a logical sequence. This will be decided by the following:

- The urgency of the items
- The length of the time each item will take

Try not to waste the first hour on trivialities and then run out of time for really important issues.

Timing

There are four aspects to timing.

1   The Actual Time Available for the Meeting

This will be determined by the other activities in the working day as well as the commitments of those participating in the meeting. A very lengthy meeting tends to be demotivating, especially if little progress is made. Short meetings encourage people to keep to the point. Frequent meetings (for example, weekly interdepartmental meetings) usually need less time than the less frequent ones (for example, quarterly management reviews).

2   The Time of Day

Most people will have attended the dreaded Friday afternoon meeting when others are concentrating on getting home for the weekend and not on the meeting. The actual time of day needs to be thought out carefully so as to cause the least disruption to all concerned. If one wants a short meeting, start it just before lunch.

3   *The Amount of Time to Spend on Each Item*

This will be determined to some extent by the amount of time available for the whole meeting. Some items may have to be left off the agenda altogether and discussed at a separate meeting. It is a good idea to work out a rough plan of how much time is available for each item individually. This may also be put on the agenda itself to act as guidance for each member. It is useful when asking for a contribution to the agenda to ask how much time will be needed for each item.

4   *Start on Time and Finish on Time*

Starting on time, even if a participant is late, shows that one means business, and it will encourage punctual attendance. Finishing on time encourages brief, meaningful contributions and also lets participants get on with their other tasks.

Give Guidance

Make sure each topic is described in enough detail so that members may come fully prepared.

Circulate

Circulate the agenda in a timely fashion before the meeting. Remember to attach any relevant reports or written briefs to the agenda and ensure participants receive them in advance.

CHECKLIST 2: THE MINUTES

There are two aspects to producing minutes: taking the notes and writing up the minutes/action notes

Taking Notes

i.   Listen carefully throughout the meeting, in particular, for the leader's interim and final summaries for guidance. If necessary, ask the leader for a summary during the meeting if implications of the discussion are not clear.

ii.   Take brief, relevant notes under each agenda item. Leave space between items in case the discussion returns to an earlier point. This saves time on sorting out afterward which notes are relevant to which points. Record the main points raised (by the importance of contribution, not length) and record the action to be taken by whom, and, if relevant, by when.

iii.   Check that notes are satisfactory after the meeting. If information is missing or vague, check with the relevant member.

Writing the Minutes/Action Notes

i.   Draft the minutes as soon as possible after the meeting.

ii. Keep all notes until the minutes have been approved (this is usually done at the next meeting).

iii. Be selective—be accurate, brief, and clear with the writing.

iv. Specify any action required in a separate column, indicating the name of the person concerned and, if appropriate, the target date.

v.   Structure the minutes so they follow the structure of the agenda and avoid complicated numbering systems.

vi. Circulate the minutes as soon as possible after the meeting, having first checked them with the leader.

Appendix B

EFFECTIVE PRESENTATIONS

It is often necessary to make a presentation on a topic in a seminar-like format using software applications, such as PowerPoint. Specifically, in the business, service, and industrial settings an oral presentation may be required for many reasons including the following:

- Proposal for a new idea
- Budget and financial report on expenditure and income
- Progress report on a project

An engineer in the role of an industrial engineer/manager is likely to be frequently in a position that requires a presentation. An accurate and well-prepared presentation goes a long way in convincing the audience and gaining support for a new endeavor, for instance. As a complement to Appendix A, we highlight some key points for preparing an effective presentation.

PREPAREDNESS

1. Prepare the required material ahead of time.
2. Check your draft carefully at least a couple of times.
3. Check the conditions, availability and type of audiovisual equipment and other materials ahead of time.

4. Have a second copy of your presentation available, in a medium suitable for the audiovisual equipment used, if the need arises.

5. Rehearse your presentation to control timing. If possible, try to rehearse your speech out loud. This will give you a better indication of the timing than whispering to self.

6. Arrive at the seminar location five minutes ahead of schedule.

SPEAKING STYLE

1. Relax and be in control.

2. Speak loudly and clearly; do not shout.

3. Face the audience; look at everybody in the room and not a particular section.

QUALITY OF THE PROJECTED TEXT

1. Use a proper font size.

2. Use short/simple equations, and not too many.

3. Use headings and subheadings.

4. Use short sentencelike descriptions, and paraphrase them in your speech.

5. Use clear drawings and plots with legible captions.

6. Use clear photographs. There is no need to include dark or unclear images and then apologize for their inferior quality. The same applies to any audio-video clip.

QUALITY OF PRESENTATION CONTENT

1. Introduction: succinctly highlight what it is all about.

2. Justify the relevance of the work.

3. Be sure of the accuracy and reliability of the material.

4. State the position of the material in the context of existing knowledge, that is, *references*:

   (a) Do not forget them.

(b)   Do not just list them!

(c)   Briefly, discuss their relevance, shortcomings, and so on.

5   Clearly explain the theory/idea proposed.

6.   Give key information about the approach/methods/equipment/experiments used.

7.   Explain the physics/mathematical treatment/assumptions considered.

8.   Highlight the expected or achieved results and contributions made.

9.   Discuss the results (if available). The discussion should not be limited to a plain explanation of numbers and graphs. A discussion highlights the reasons for a particular result obtained, or how it can be affected.

10. End with a solid conclusion. A conclusion is not a summary; it explains what is learned and what is contributed.

OTHER QUALITIES

1.   The content of the presentation must be suitable for the experts as well as for the novice in the field. This is, of course, a tough task to accomplish.

2.   Respond to questions clearly, giving short, to-the-point answers so as to provide the opportunity for more people to ask questions.

3.   In your answers, treat the audience with respect!

BIBLIOGRAPHY

Aft, Lawrence S. *Fundamentals of Industrial Quality Control*, CRC Press, 1988.

Allen, Theodore T. *Introduction to Engineering Statistics and Six Sigma: Statistical Quality Control and Design of Experiments and Systems*, Springer, 2006.

Ashore, S. *Sequencing Theory*, Springer-Verlag, 1972.

Baker, Kenneth R. *Elements of Sequencing and Scheduling*, Kenneth R. Baker, 1993.

Basu, Ron.*Implementing Six Sigma and Lean—A Practical Guide to Tools and Techniques*, Elsevier, 2009.

Bedworth, David D., and Bailey, James E. *Integrated Production Systems*, Wiley, 1987.

Blanchard, Benjamin S. *Logistics Engineering and Management*, Prentice-Hall, 1986.

Bridges, William.*The Character of Organization*, Consulting Psychologists Press, 1992.

Chandra, M. Jeya.*Statistical Quality Control*, CRC Press, 2001.

Chase, Richard B., and Aquilano, Nicholas J. *Production and Operations Management*, Irwin, 1995.

Conway, Richard W., Maxwell, William L., and Miller, Louis W.*Theory of Scheduling*, Addison-Wesley, 1976.

Elbert, Mike.*Lean Production for the Small Company*, CRC Press, 2013.

Elsayed, Elsayed A., and Boucher, Thomas O. *Analysis and Control of Production Systems*, Prentice-Hall, 1985.

Evans, James R. *Applied Production and Operations Management*, West Publishing, 1993.

Evans, James R. *Production/Operations Management—Quality, Performance, and Value*, West Publishing, 1993.

Evans, James R., and Lindsay, William M. *The Management and Control of Quality*, West Publishing, 1993.

Field, S. W., and Swift, K. G. *Effecting a Quality Change*, Arnold, 1996.

Francis, R. L. *Geometrical Solution Procedure for Rectilinear Distance, Minimax Location Problem*, AIIE Transactions, 1972.

Francis, R. L., and White, John A. *Facility Layout and Location—An Analytical Approach*, Prentice-Hall, 1974.

French, S. *Sequencing and Scheduling—An Introduction to the Mathematics of the Job-Shop*, Ellis Horwood, 1986.

George, Mike, Rowlands, Dave, and Kastle, Bill.*What Is Lean Six Sigma*, McGraw-Hill, 2004.

Grant Eugene L. *Statistical Quality Control*, McGraw-Hill, 1996.

Groover, Mikell P. *Automation, Production Systems, and Computer-Aided Manufacturing*, Prentice-Hall, 1980.

Groover, Mikell P. *Automation, Production Systems, and Computer-Integrated Manufacturing*, Prentice-Hall, 1980.

Handfield, Robert B. *Introduction to Supply Chain Management*, Prentice-Hall 1999.

Hatch, Mary Jo, with Cunliffe, Ann, L. *Organization Theory—Modern, Symbolic, and Postmodern Perspectives*, Oxford University Press, 2013.

Henley, Ernest J., and Williams, R. A. *Graph Theory in Modern Engineering*, Academic Press, 1973.

Herzog, Donald R. *Industrial Engineering Methods and Controls*, Reston Publishing, 1985.

Hirano, Hiroyuki. *JIT Factory Revolution—A Pictorial Guide to Factory Design of the Future*, Productivity Press, 1988.

Jacobs, F. Robert, and Chase, Richard.*Operations and Supply Chain Management—The Core*, McGraw-Hill Ryerson, 2010.

Jardine, A. K. S. *Maintenance, Replacement and Reliability*, Pitman Publishing, 1973.

Jardine, A. K. S., and Tsang, Albert H. C. *Maintenance, Replacement and Reliability*, CRC Press, 2006.

Joglekar, Anand M. *Statistical Methods for Six Sigma—In R&D and Manufacturing*, Wiley, 2003.

Johnson, Lynwood A., and Montgomery, Douglas C. *Operations Research in Production Planning, Scheduling, and Inventory Control*, Wiley, 1974.

Jozefowska, Joanna.*Just-in-Time Scheduling—Models and Algorithms for Computer and Manufacturing Systems*, Springer, 2007.

Jugulum, Rajesh, and Samuel, Philip.*Design for Lean Six Sigma*, Wiley, 2008.

Kilduff, Martin, and Krackhardt, David.*Interpersonal Networks in Organizations—Cognition, Personality, Dynamics, and Culture*, Cambridge University Press, 2008.

Kline, Theresa J. B. *Teams That Lead*, Lawrence Erlbaum Associates, 2003.

Kuhn, H. W. *On the Pair of Dual Non-Linear Problems, Non-Linear Programming, Chapter 3*, J. Abadie, Ed., Wiley, 1967.

Kusiak, Andrew, Editor.*Flexible Manufacturing Systems: Methods and Studies*, Elsevier Science Publishers, 1986.

Kutz, Myer.*Environmentally Conscious Manufacturing*, Wiley, 2007.

Lee, Sang M., Moore, Laurence J., and Tayler Bernard W. *Management Science*, Wm. C. Brown Company Publishers, 1981.

Lenz, John E. *Flexible Manufacturing—Benefits for Low-Inventory Factory*, Marcel Dekker, 1989.

Littlechils, S. C., Editor.*Operational Research for Managers*, Philip Allan Publishers, 1977.

Luggen, William W. *Flexible Manufacturing Cells and Systems*, Prentice-Hall, 1991.

Mackie, Dan.*Engineering Management of Capital Projects—A Practical Guide*, McGraw-Hill, 1984.

Makower, M. S., and Williamson, E. *Operational Research*, Hodder and Stoughton, 1975.

Mayer, Raymond R.*Production and Operations Management*, McGraw-Hill, 1982.

Miller, David M., and Schmidt, J. W. *Industrial Engineering and Operations Research*, Wiley, 1984.

Montgomery, Douglas C. *Introduction to Statistical Quality Control*, Wiley, 2013.

Myerson, Paul A. *Supply Chain Management Made Easy: Methods and Applications for Planning, Operations, Integration, Control and Improvement, and Network Design*, Pearson Education, 2015.

Opitz, H. *A Classification System to Describe Workpieces*, Pergamon Press, 1970.

Page, E. *Queueing Theory in OR*, Butterworths, 1972.

Papadopoulos, H. T., Heavey, C., and Browne, J. *Queueing Theory in Manufacturing Systems Analysis and Design*, Chapman & Hall, 1993.

Pinedo, Michael.*Scheduling—Theory, Algorithms, and Systems*, Prentice-Hall, 1995.

Puccio, Gerard J., Mance, Marie, and Murdock, Mary C. *Creative Leadership—Skills That Drive Change*, Sage Publications, 2011.

Pyzdek, Thomas.*What Every Engineer Should Know about Quality Control*, Marcel Dekker, 1989.

Robson, George D. *Continuous Process Improvement—Simplifying Work Flow Systems*, Free Press, 1991.

Ryan, Thomas P. *Statistical Methods for Quality Improvement*, Wiley, 2011.

Santos, Javier, Wysk, Richard A., and Torres, Jose M. *Improving Production with Lean Thinking*, Wiley, 2006.

Sarkis, Joseph.*Green Supply Chain Management*, ASME Press, 2014.

Schroeder, Roger G. *Operations Management*, McGraw-Hill, 1985.

Shooman, M. L. *Probabilistic Reliability—An Engineering Approach*, McGraw-Hill, 1968.

Singh, Chanan, and Billinton, Roy.*System Reliability Modelling and Evaluation*, Hutchinson Publishers, 1977.

Stapenhurst, Tim.*Mastering Statistical Process Control—A Handbook for Performance Improvement Using Cases*, Elsevier, 2005.

Stebbing, Lionel.*Quality Assurance—The Route to Efficiency and Competitiveness*, Elis Horwood, 1993.

Sule, Dileep R. *Production Planning and Industrial Scheduling*, CRC Press, 2008.

Taguchi, Genichi.*Introduction to Quality Engineering—Designing Quality into Products and Processes*, Asian Productivity Organization, 1989.

Taguchi, Genichi, Elsayed, A., and Hsiang, Thomas.*Quality Engineering in Production Systems*, McGraw-Hill, 1989.

T'kindt, Vincent, and Billaut, Jean-Charles.*Multicriteria Scheduling—Theory, Models and Algorithms*, Springer, 2002.

Tompkins, James A. *Facilities Planning*, Wiley, 1984.

Tompkins, James A., White, John, A., Bozer, Yavuz A., and Tanchoco, J. M. A. *Facilities Planning*, Wiley, 2003.

Turner, Wayne C., Mize, Joe H., and Case, Kenneth E.*Introduction to Industrial and Systems Engineering*, Prentice-Hall, 1987.

Wood, Douglas C., Editor.*Principles of Quality Costs—Financial Measures for Strategic Implementation of Quality Management*, ASQ Quality Press, 2013.

Wu, B. *Manufacturing Systems Design and Analysis*, Chapman & Hall, 1992.

INDEX